# CAMBRIDGE LIBRARY COLLECTION

*Books of enduring scholarly value*

## Life Sciences

Until the nineteenth century, the various subjects now known as the life sciences were regarded either as arcane studies which had little impact on ordinary daily life, or as a genteel hobby for the leisured classes. The increasing academic rigour and systematisation brought to the study of botany, zoology and other disciplines, and their adoption in university curricula, are reflected in the books reissued in this series.

## Fungi

Dame Helen Gwynne-Vaughan (1879–1967) was a prominent British mycologist, specialising in the sexual process of fungi. In 1909 she was appointed Head of the Department of Botany at Birkbeck College, becoming Professor of Botany when Birkbeck College joined the University of London in 1920. This volume was first published in 1922 as part of the Cambridge Botanical Handbooks series. The introduction provides a detailed description of the structure, sexual reproduction, parasitism and symbiosis of all fungi, with subsequent chapters describing fully the morphology and reproduction of genera within the phylum ascomycetes and the orders ustilaginales and uredinales on which Gwynne-Vaughan based her research. Illustrations and a bibliography accompany each chapter. This volume provides an insight into the study of mycology in the early twentieth century, before technological advances in the field of cytology revolutionised the discipline.

Cambridge University Press has long been a pioneer in the reissuing of out-of-print titles from its own backlist, producing digital reprints of books that are still sought after by scholars and students but could not be reprinted economically using traditional technology. The Cambridge Library Collection extends this activity to a wider range of books which are still of importance to researchers and professionals, either for the source material they contain, or as landmarks in the history of their academic discipline.

Drawing from the world-renowned collections in the Cambridge University Library, and guided by the advice of experts in each subject area, Cambridge University Press is using state-of-the-art scanning machines in its own Printing House to capture the content of each book selected for inclusion. The files are processed to give a consistently clear, crisp image, and the books finished to the high quality standard for which the Press is recognised around the world. The latest print-on-demand technology ensures that the books will remain available indefinitely, and that orders for single or multiple copies can quickly be supplied.

The Cambridge Library Collection will bring back to life books of enduring scholarly value (including out-of-copyright works originally issued by other publishers) across a wide range of disciplines in the humanities and social sciences and in science and technology.

# Fungi

*Ascomycetes, Ustilaginales, Uredinales*

HELEN GWYNNE-VAUGHAN

CAMBRIDGE UNIVERSITY PRESS

Cambridge, New York, Melbourne, Madrid, Cape Town, Singapore,
São Paolo, Delhi, Dubai, Tokyo, Mexico City

Published in the United States of America by Cambridge University Press, New York

www.cambridge.org
Information on this title: www.cambridge.org/9781108013215



This edition first published 1922
This digitally printed version 2010

ISBN 978-1-108-01321-5 Paperback

Cambridge Botanical Handbooks

Edited by A. C. SEWARD and A. G. TANSLEY

# FUNGI

## ASCOMYCETES, USTILAGINALES, UREDINALES

CAMBRIDGE UNIVERSITY PRESS

C. F. CLAY, MANAGER

LONDON : FETTER LANE, E.C. 4

LONDON : H. K. LEWIS AND CO., LTD.,
136, Gower Street, W.C. 1

LONDON : WHELDON & WESLEY, LTD.,
28, Essex Street, Strand, W.C. 2

NEW YORK : THE MACMILLAN CO.

BOMBAY }
CALCUTTA } MACMILLAN AND CO., LTD.
MADRAS }

TORONTO : THE MACMILLAN CO. OF CANADA, LTD.

TOKYO : MARUZEN-KABUSHIKI-KAISHA

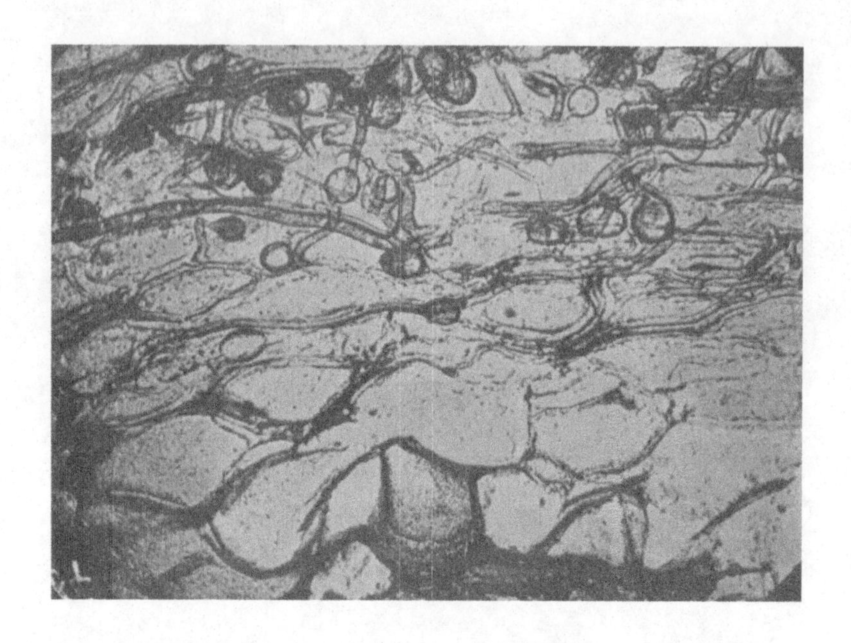

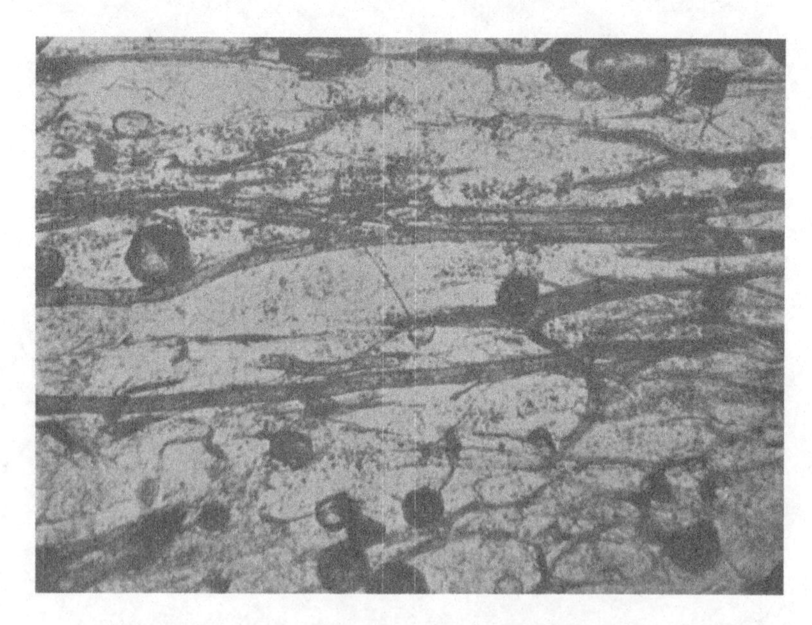

**PALEOMYCES ASTEROXYLI**

from the Old Red Sandstone, Muir of Rhynie, Aberdeenshire, × 100; after Kidston and Lang

# FUNGI

## ASCOMYCETES, USTILAGINALES, UREDINALES

BY

DAME HELEN GWYNNE-VAUGHAN,
(FORMERLY H. C. I. FRASER)
D.B.E., LL.D., D.Sc., F.L.S.
PROFESSOR OF BOTANY IN THE UNIVERSITY OF LONDON
AND
HEAD OF THE DEPARTMENT OF BOTANY, BIRKBECK COLLEGE

CAMBRIDGE
AT THE UNIVERSITY PRESS
1922

# PREFACE

IT is impossible to study the Fungi without being impressed by the undiminished value of much of the older work, and especially of that of de Bary, or without recognizing the soundness of his general and very many of his particular conclusions in the light of subsequent investigation. While I have tried to give something approaching adequate references to recent literature, I have thought it superfluous to name the more general works of those earlier authors who are quoted in de Bary's *Comparative Morphology of the Fungi, Mycetozoa and Bacteria* and elsewhere. By such investigations the foundations of modern mycology have been laid and their discoveries have passed into the groundwork of our knowledge.

The intention of the following pages is to present the fungus as a living individual: the scope is mainly morphological, but, in dealing with objects so minute, morphology passes insensibly into cytology. The introduction deals with fungi in general; the special part of this volume is limited to the consideration of the Ascomycetes, Ustilaginales and Uredinales. The manuscript was completed early in 1917, but an endeavour has been made to bring it up to date.

The majority of the illustrations are drawn from published researches, and I have to thank those authors who have given me permission to copy their figures. Illustrations, the source of which is not stated, are original. In the case of original figures the magnification and the authority for the species are given; this has also been done in other cases whenever the information was available.

I am grateful to many past and present students for specimens and information; to Miss W. Page for figure 112; to Miss H. Tayler for reading proofs; to Mr Charles Dobb for valuable help in the preparation of figures; to Mr E. S. Salmon for advice on the section dealing with specialization of parasitism; and especially to my friends Miss E. J. Welsford and Mr J. Ramsbottom for assistance in a number of ways.

The earlier written parts of the book, and consequently the whole, owe much to the unfailing interest and wise criticism of my husband.

H. C. I. GWYNNE-VAUGHAN.

LONDON,
*September*, 1921

# CONTENTS

[1] Books and papers dealing with individual fungi or groups of fungi are cited in the Bibliography at the end of the relevant section or in foot-notes in the text.

# CHAPTER I

## INTRODUCTION

**The Fungi** are parasitic or saprophytic Thallophyta entirely destitute of chlorophyll, and possessing in the very large majority of cases a vegetative portion, the **mycelium**, made up of filaments or **hyphae**. The group is a very ancient one, the earliest known undoubted fungi occurring among the remains of *Rhynia* and *Hornia* in the Old Red Sandstone of the Muir of Rhynie, Aberdeenshire. This material consists of aseptate hyphae and vesicles which doubtless served the purpose of reproduction (frontispiece)[1].

Fungal hyphae may be non-septate and coenocytic, or they may undergo transverse septation, in which case their constituent cells are either uninucleate or multinucleate. Any division other than transverse is extremely rare; it occurs, for example, in the development of certain multicellular (muriform) spores (fig. 1), and in the initiation of the perithecium in *Strickeria* and of the pycnidium in *Pleospora* and *Phoma*[2].

As a rule the hyphae are richly branched; they elongate by apical growth and usually spread loosely through the substratum; in certain cases, especially in relation to the fructifications of the higher forms, they become woven into a dense mass which gives in section the appearance of a tissue, and is therefore described as **pseudoparenchymatous**; when fructifications are embedded in such a mass it is termed a **stroma**; a similar weft of hyphae sometimes give rise to root-like strands of which the best example is the so-called **rhizomorph** of *Armillaria mellea*, or to a compact resting body or **sclerotium** the outer cells of which are modified to form a thick-walled rind, protecting the vegetative mycelium against desiccation.

Frequent anastomoses take place between hyphae, either by means of short branches forming loops, bridges or **H-pieces**, or by means of so-called **clamp-connections** which join adjacent cells; such arrangements facilitate the passage of food and may, in certain cases, become sufficiently numerous to form a net-work.

The mycelium begins its development as a **germ-tube** put out from one of the numerous types of fungal **spore**. Where the spore wall is very thin the wall of the germ-tube may be continuous with it (zoospores), but in the majority of cases the wall of the germ-tube is continuous only with the

[1] Kidston, R. and Lang, W. H. On Old Red Sandstone Plants showing Structure from the Rhynie Chert Bed, Aberdeenshire, *Trans. Roy. Soc. Ed.* 1921.

[2] Kempton, F. E. Origin and Development of the Pycnidium, *Bot. Gaz.* 1919, lxviii, p. 233.

inner layer of the spore wall. In such cases one or more germ-tubes may break through the wall of the spore at spots not previously recognizable, or they may find an exit through special pits or **germ-pores** formed during the development of the spore. The germ-tube elongates and receives the contents of the spore.

In cases where a mycelium is not developed the plant body consists entirely of reproductive structures (Yeast, Archimycetes).

The typical fungal protoplast consists of a mass of granular or reticulate cytoplasm, which in the older regions leaves a vacuole in the centre of the cell or filament; the nucleus, where its size has permitted of detailed investigation, has a structure quite similar to that of other plants and animals, and usually divides by mitosis, showing a well-marked spindle with centrosomes and asters. The development of the spindle is extranuclear in certain Uredinales. One or more nucleoli are commonly present and are thrown out into the cytoplasm during karyokinesis. The extrusion of chromatin bodies has been described in *Helvella crispa*.

The cell wall consists of cellulose; often a special variety known as fungus cellulose is present. The storage materials include amylo-dextrin or soluble starch, amyloid, a reserve-cellulose, both of which turn blue with iodine; oil, glycogen, and various protein substances. The protoplasm gives rise also to a number of ferments which not only enable the plant to deal with its food materials, but bring into solution the walls of the host cells, and so make possible the penetration of parasitic hyphae.

**Sexual reproduction** among the fungi takes place by the union of two uninucleate or multinucleate cells which may be similar in structure and behaviour, or may be differentiated as an antheridium and an oogonium. Each of these organs contains one or more distinct gametes, or else a number of gametes which do not become rounded off from one another or separated from the wall of the parent cell, but are indicated by separate nuclei lying in an undifferentiated mass of cytoplasm. To organs of the latter type the term **coenogamete** is sometimes applied in recognition of their multinucleate character; it is, however, inappropriate, since they are not gametes, but gametangia. In the vast majority of fungi free swimming gametes are not developed; the sole exceptions are found in the genus *Monoblepharis*, where uniciliate or biciliate spermatozoids are set free and swim to the female organ.

A state of affairs in which the antheridium as a whole must grow or be carried to the oogonium involves a risk that normal fusion will fail to occur, while at the same time the presence of multinucleate sexual organs and of vegetative cells between which anastomoses readily occur offers considerable opportunities for some form of "reduced" fertilization. The replacement of normal fertilization by the fusion of two female or two vegetative nuclei, or of a female and a vegetative nucleus, is very common among fungi, and a complete disappearance of even this reminiscence of a sexual process is by no means rare. It has been suggested that the variety of food material which fungi as parasites and saprophytes obtain from their substratum may make the stimulus of fertilization less important, and it is possible also that among these plants competition is less severe than among holophyta or holozoa. At any rate the group shows a progressive disappearance of normal sexuality.

The sexual fusion or its equivalent is followed in all investigated cases by a reducing division or meiotic phase, so that, as in other plants or animals, the number of chromosomes is doubled in fertilization and subsequently halved in meiosis, and diploid[1] and haploid phases follow one another.

The meiotic phase is usually associated with spore-formation which, in many of the lower fungi, takes place on the germination of the zygote. In a much greater number of cases a period of vegetative development intervenes between the association of the nuclei in fertilization or otherwise and chromosome reduction, and we have a well-marked alternation of generations in which a haploid gametophyte bears the sexual cells or their equivalent, and a diploid sporophyte gives rise to spores which in turn constitute the first stage of a new gametophytic generation. It is not at all uncommon to find several sporophytes arising from a single gametophyte, and the gametophytic mycelium frequently sends out branches which grow around and protect the sexual cells and their products. Where fertilization or any equivalent process has wholly disappeared we may expect to find a similar morphological alternation of generations, though without the corresponding cytological changes; but in some cases, as in the large group of *Fungi imperfecti*, a sporophyte is no longer developed, or at any rate has not been identified.

**Spores** and **Spore mother-cells.** In the higher fungi the characteristic spores of the sporophyte, with the development of which meiosis is definitely associated, may be produced either *endogenously* as **ascopores** in a mother-cell of definitely restricted size termed an **ascus**, or *exogenously* as **basidiospores** on the exterior of a cell or row of cells known as a **basidium**. The asci or basidia are frequently arranged in parallel series forming a fertile layer or **hymenium** sometimes of considerable extent. They arise from a **sub-hymenial** layer immediately below the hymenium, and among them are interpolated elongated vegetative cells or **paraphyses**, which are probably concerned in their nutrition and perhaps assist spore dispersal by keeping the mother-cells separate. The ascus and basidium and their products have long been recognized as essential features in classification.

[1] The diploid unit may be defined as a protoplast the nuclear content of which includes the double number of chromosomes.

In the lower fungi, spore formation may be associated with the meiotic phase, but the spores produced resemble those concerned in the accessory methods of reproduction.

**Accessory spores.** The **accessory** or non-sexual methods of reproduction have no relation to any sexual process either normal or reduced, and therefore no significance in the alternation of generations; they are devices for rapid multiplication comparable with the gemmae in *Marchantia* or the arrangements for vegetative propagation in higher plants. The spores concerned may be borne either on the sporophyte (rusts, etc.) or, as in the majority of cases, on the gametophyte.

In many of the lower fungi **zoospores** are developed in spherical, ovoid or tubular **zoosporangia**; this is the case especially in aquatic forms. In relation to the change from aquatic to subaerial conditions the contents of the sporangium may come to be shed as **walled non-motile spores**, or the sporangium may itself be set free without division of its contents. Such a structure, borne externally on its parent hypha, is termed a **conidium**, and is the characteristic accessory reproductive unit of the fungi. In the large majority of cases the conidium germinates by means of a germ-tube, but where the fungus has not completely abandoned its aquatic habit the conidium, if it falls in wet conditions, may give rise to zoospores either internally or in a vesicle borne on a short hypha. The conidia are developed either singly or in groups on **conidiophores**; these may be free, they may be gathered into a sheaf or **coremium**, or they may be formed inside a special flask-shaped receptacle known as a **pycnidium**; they show an almost endless variety in form and arrangement.

A less common reproductive cell is the **chlamydospore**; these are borne either singly or in chains in the course of the ordinary vegetative hyphae or at the ends of special branches; they are characterized, as their name implies, by an exceptionally thick wall.

In certain species and under certain conditions whole hyphae may break up into series of separate cylindrical cells or spores. Such a spore is termed an **oidium**. Oidium-formation appears to be a rapid and efficient method of multiplication and is the only one found in the fungi of such diseases as favus (*Achorion Schoenleinii*), pityriasis versicolor (*Microsporon furfur*) and thrush (*Monilia albicans*).

In these cases attempts to cultivate any more characteristic fructification have failed, and the fungus cannot therefore be assigned to any particular group.

**Morphology of the spore.** The individual spore whether belonging to the principal or accessory fructification is, when first formed, a **hyaline**, colourless cell; in the course of development it may divide to produce a row or a mass of cells and in the latter case is described as **muriform**;

it may also become variously coloured. In the large majority of cases the spore is enclosed by a double wall consisting of a delicate endospore and an epispore which may be smooth or variously sculptured; it may develop small projections and is then said to be warted or **verrucose**, or it may be **reticulate**, exhibiting a number of more or less regular polygonal depressions between which anastomosing ridges are present.

Many conidia and other thin walled fungal spores possess the power, in suitable media, of **budding** or **sprouting**; giving rise, that is to say, to new cells as simple lateral outgrowths which are soon nipped off. This method of propagation is shown in the conidia of the yeasts, in some of which it has wholly superseded the development of a mycelium. Ascospores are found to bud in the Exoascaceae, and basidiospores in the Ustilaginales.

**Classification.** The fungi are divided into three great groups according to the septation of their mycelium, and the characters of their principal spores.

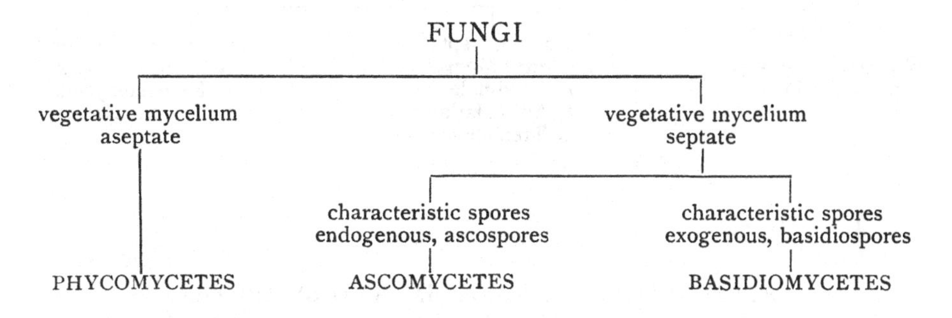


They may be further subdivided as follows:

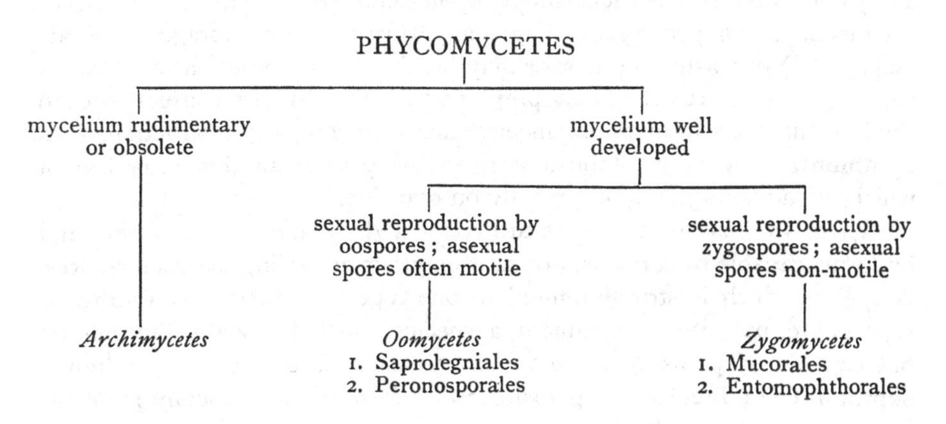

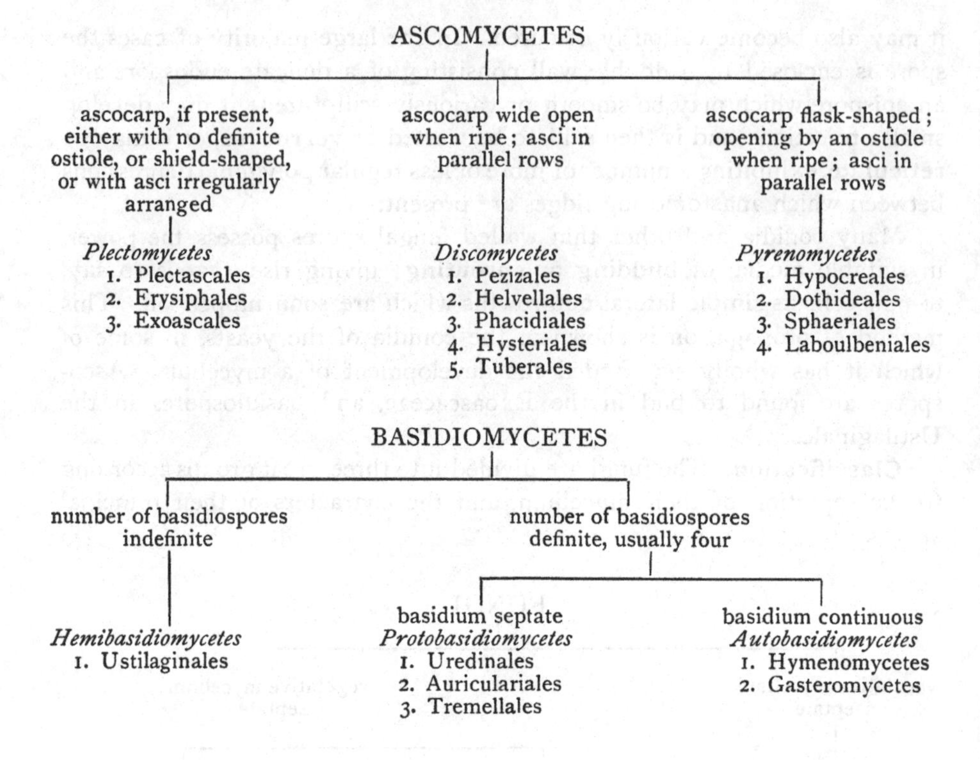


## SAPROPHYTISM, PARASITISM AND SYMBIOSIS

Since fungi under no circumstances possess chlorophyll, they are necessarily dependent for their food supply upon some sort of relation with another organism. As **saprophytes** they may utilize organic storage materials (sugar, etc.) or waste products, or may break up dead tissues as a source of supply; as **parasites** they may prey upon living cells with consequences to the host that vary from trifling inconvenience to complete destruction, or as **symbionts** they may establish a relationship with another organism in which the advantages are not wholly on one side.

These various arrangements are connected by intermediate forms, and by forms capable of parasitism or saprophytism according to circumstances. A species which is strictly limited to one type of nutrition is an **obligate** saprophyte, parasite, or symbiont, a species which is usually saprophytic but capable of parasitic existence on occasion, is described as a **hemi-saprophyte** or **facultative** parasite, and a form which is usually parasitic, but sometimes saprophytic as a **hemi-parasite**, or **facultative** saprophyte.

## SAPROPHYTISM

According to our present knowledge the large majority of fungi are saprophytic; a considerable proportion of forms in each of the great groups and especially a very large number of the Basidiomycetes obtain their nutrition in this way.

**On Wood.** Ascomycetes and Basidiomycetes are important agents in the breaking up of wood; their hyphae absorb the starch and protoplasm of the unaltered cells of the wood and medullary rays and penetrate into the fibres, vessels and tracheids, either passing through the pits and especially the bordered pits, or penetrating the walls. They act upon the walls so that these become delignified and give characteristic cellulose reactions and the middle lamella is dissolved. The enzyme responsible for this change was first isolated by Czapek in the case of *Merulius lacrymans*, the fungus of dry rot. Its action seems to spread in a plane parallel to the surface of the wall either from the pits, which thus become much enlarged, or from the delicate passages left by the protoplasmic connections which originally traversed the walls of the young wood elements. The whole mass of wood loses weight and may reach the easily broken and almost powdery condition known as touch-wood. In this way considerable damage may be done to timbers (dry rot, *Merulius lacrymans*), paving blocks (*Lentinus lepideus*), etc., but also considerable advantage may ensue from the restoration to the soil of the material of fallen tree trunks, twigs and branches.

The part played by the higher fungi is here specially important as almost the only other agents of destruction of lignified tissues seem to be certain molluscs and crustacea which act by boring into the wood.

**On Soil.** Yeasts and filamentous fungi are abundant in woodland soils and they are also of frequent occurrence in cultivated soil; the microscopic forms show a remarkable similarity in different localities; even in Europe and America the same genera and often the same species are obtained; in culture there is a regular succession of forms, first the Mucoraceae, then *Penicillium* and *Eurotium* and later the black and brown Hyphomycetes. A large number of Basidiomycetes also develop in the soil.

The fungi of the soil utilize the sugar, starch, pectose and hemi-cellulose which are returned to the ground in dead plants and plant organs, and, in common with certain bacteria, they act upon cellulose, breaking it up into soluble substances and humus. In several cases evidence has been brought forward that some of these fungi are capable of assimilating free nitrogen but negative results have also been very common.

The activity of these fungi is well exemplified by the "fairy-rings" of dark green grass often seen in poor pastures. The soil just outside the ring is rich in the mycelium of one or two common fungi; few hyphae are found under the ring itself and none in the area enclosed by it. The ring is dependent upon the growth of the fungus which spreads outwards in all directions from the centre, the mycelium dying off as the food materials in the soil are exhausted; in the transition region, where the fungal hyphae themselves are disintegrating, the soil is in high condition, it contains organic residues recently formed and capable of rapid change and the grass is especially luxuriant; the ring accordingly is just inside the region of maximum fungal activity.

A certain number of fungi belonging to the Tuberales, Elaphomycetaceae, Terfeziaceae and Hymenogasteraceae are completely subterranean or **hypogeal** in their development. They produce closed fructifications protected by a stout wall of interwoven hyphae. As the spores approach maturity the fructifications develop a strong scent, varying much in character and from the human standpoint either pleasant or disagreeable, which serves to attract animals and especially rodents. The fructification is eaten and the spores pass uninjured through the alimentary canal, and are thus distributed. The truffles (*Tuber spp.*) are the best known of these forms.

**Coprophilous Fungi.** Fungi feeding on organic remains in the soil often benefit by the presence of natural manures and incidentally help to break up these substances so that they become available for the higher plants.

From such fungi it is no great transition to the extensive **coprophilous** flora of which the habitat is the dung of various animals and especially of herbivorous species. In addition to the rich nitrogenous food supply which these fungi obtain, the presence of cellulose in the straw and other vegetable debris in the dung is an important factor in their nutrition. This is well shown by the fact that many coprophilous species fail to fruit in artificial culture of dung decoction and agar, unless they are provided with cellulose. Cotton wool or pieces of filter paper laid on the substratum admirably serve this purpose; the latter are soon broken into small flocculent scraps. In nature Zygomycetes, Ascomycetes and Basidiomycetes succeed each other in fairly regular order and, speaking generally, show very similar adaptations to their habitat. In many Ascomycetes (Ascobolaceae, Sordariaceae) the spores are surrounded by mucilage and form together a projectile which owing to its weight can be shot to a much greater distance than would be possible for single spores. The sudden ejection of the spore mass seems to depend on the absorption of water by the mucilaginous contents of the ascus. After ejection the mass dries up and becomes firmly attached to the substratum on which it has fallen. In the same way the spores in the sporangium of *Pilobolus* are surrounded by a gelatinous envelope which swells in the presence of water and bursts the sporangium wall, so that the whole sporangium is shot off as a single mass and adheres by means of the gelatinous layer to the body against which it strikes.

The grass surrounding the dung receives an ample supply of spores and spore masses; later, if it is eaten by some herbivorous animal, the spores pass uninjured through the alimentary canal and germinate while still in the intestine or on being ejected with the dung. In some cases the wall of the spore seems to have become so effectually adapted to resist injury during its passage through the animal that it is incapable of either stretching or cracking as a preliminary to germination except after digestion or some other special treatment.

Thus de Bary succeeded in germinating spores of species *Pilobolus* and *Mucor* in water and those of *Sordaria* and *Coprinus* in nutrient solution, by exposing them to a temperature between 35° and 40° C.; Massee and Salmon obtained germination in the spores of *Ascobolus perplexans*, and *A. glaber*, after about 20 hours in either tap water or dung decoction, at a temperature of about 27° C.; I found that spores of *Lachnea stercorea* germinated in an alkaline medium (preferably dung decoction) after incubation for several hours at 38° C. (the temperature of the body of the cow), and Welsford succeeded by the same means in germinating the spores of *Ascobolus furfuraceus*, and Cutting those of *Ascophanus carneus*. Ramlow, however, describes the germination of spores of the last named species at room temperature in twenty-four hours. Further, Dodge was able to germinate the spores of several Ascobolaceae on dung agar by exposing them for 5 or 10 minutes to a temperature of 50° to 70° C., and Ramlow germinated those of *Ascobolus immersus* by sowing them on agar which was still hot after sterilization. In certain cases the action of direct sunlight was found by Dodge sufficient to induce a moderate percentage of germinations and to raise the temperature of the liquid containing the spores to about 50° C. in half an hour.

Finally, as I am informed by Mr Ramsbottom, germination may be induced mechanically by cracking or breaking the epispore, for example by rubbing the spores between two coverslips, so that all the above methods are a mere variety of means towards this end.

In the case of *Lachnea stercorea*, spores incubated for 18 to 40 hours, either in a succession of digestive fluids, or in dung extract, only germinated approximately 50 hours after the beginning of the experiment. It is probable that they do so in nature about two days after being swallowed. In *Ascophanus carneus*, also common on cow dung, germination is much more rapid, taking place under similar circumstances in a single night, so that the spores under natural conditions may be inferred to germinate while actually in the intestine. Further development, however, is in all cases dependent upon the fungus reaching the exterior of the mass of dung.

In *Coprinus sterquilinus*, Baden found that not only warmth and an alkaline medium (aqueous extract of horse dung), but the presence of certain bacteria also was necessary for germination. Appropriate bacteria may sometimes be a factor in the development of the ascocarp (Molliard) and there are many indications that certain fungi grow better in impure than in pure cultures. This, however, may merely indicate that in a mixed culture the waste products of one organism are used up by the others, whereas in a pure culture they tend to accumulate and inhibit growth. Thus Dodge found that spores of *Ascobolus Winteri* failed to germinate on the agar in which the parent ascocarps were growing.

In the case of *Mucor* and *Rhizopus*, Baden found that the presence of bacteria prevented the germination of the spores. It may be worth considering whether the progressive development of bacteria in the dung may be a factor in the succession of fungi which appear on it.

Another factor of some importance in the development of coprophilous and perhaps of other fungi is the action of direct sunlight. Cultures which remain sterile in a darkened room can often be induced to fruit by placing them in a sunny window; reference has already been made to the occasional action of sunlight on spores. Many coprophilous fungi are moreover positively heliotropic; this is well shown by the sporangiophores of *Pilobolus* and the perithecia of *Sordaria* and its allies. The ejection of the spores into an open space is in this way ensured.

**Fungi on Fatty Substrata.** It is probable that all or most fungi are able to utilize fats and oils; such substances are a common form of food reserve in the spores (zoospores, oospores, uredospores, etc.), in the mycelium, and especially in the sclerotia where, in the case of *Claviceps purpurea*, the proportion of fat reaches as much as 35 per cent.; in several cases the fat-splitting enzyme, lipase, has been extracted.

It is therefore not surprising that many fungi grow readily on a fatty substratum, some, such as *Empusa* and species of *Cordyceps*, on animal remains, some on other fungi and some on oil-containing fruits and seeds and on cotton, rape and other oil-cakes which are made from the waste remaining after such seeds have been crushed; they may reduce the oil content of the cake from over 10 per cent. to between 1 and 2 per cent. in two years. *Eurotium* and *Penicillium* occur on the layer of sweet oil placed over bottled fruits to prevent decomposition and together with other genera are concerned in the "ripening" of cheese. The related *Monascus heterosporus* does considerable damage in parts of Australia and New Zealand if it is allowed to get a footing on stored tallow.

**Fungi producing alcoholic Fermentation.** A number of fungi obtain nutriment from solutions consisting largely of soluble carbohydrates and they may also obtain energy by directly breaking up certain of these substances without the intervention of oxygen and with the formation of ethyl alcohol, carbon dioxide and small quantities of other substances. This reaction is due to the presence of the enzyme zymase and is known as alcoholic fermentation.

In its simplest possible form it may be represented by the equation:

$$C_6H_{12}O_6 = 2C_2H_6O + 2CO_2$$

Only certain monosaccharides with the formula $C_6H_{12}O_6$ (glucose, fructose) are capable of undergoing alcoholic fermentation; polysaccharides (cane-sugar, lactose, maltose) must undergo a preliminary hydrolysis resulting in the production of appropriate monosaccharides before alcoholic fermentation can take place.

The majority of the fungi capable of inducing these changes are yeasts (Saccharomycetaceae) acting either alone or in conjunction with appropriate bacteria; they are very commonly present as epiphytes on the skin of ripe fruits and feed on the drops of sugary solution that may be exuded or escape where the skin is broken. At other times, even in the winter, they may be found in the neighbouring soil, but they are very rarely present upon unripe fruits, presumably because any cells that happen to be carried there soon die.

A number of yeasts are made use of economically in baking, where their value depends on the formation of carbon dioxide, causing the dough to "rise," as well as in brewing and the other processes concerned with the production of alcohol. The characteristic yeast of wine which ferments glucose (or grape-sugar) is found in abundance at vintage time on the grapes and their stalks, and the cider yeasts on apples; the yeast of beer on the other hand, which acts on the sugar formed in germinating barley, is not known in the wild state.

In the production of a number of alcoholic beverages the yeast acts symbiotically with one or more bacteria; this is the case in the group of organisms included in the "ginger-beer plant"[1] which, added to commercial ginger, sugar and water, causes the formation of ginger-beer. The "plant" has the appearance of lumpy irregular masses rather like pieces of soaked tapioca or sago; its essential constituents, as Marshall Ward demonstrated in 1892, are the yeast *Saccharomyces pyreformis* and the bacterium *B. vermiforme*; the bacterium is able to utilize the products of the metabolism of the yeast, and can do so most successfully at their first formation, that is, in the neighbourhood of the living yeast cell; the yeast benefits by the removal of these substances, the accumulation of which would inhibit its development, and is able, in the presence of the bacterium (and of appropriate food materials), to continue its activity for weeks as shown by the evolution of carbon dioxide. The relation between the two organisms is thus a symbiosis in which each constituent gains by its association with the other.

A similar combination appears to exist in the materials used for two different fermentations of lactose or milk sugar which underlie the productions of the beverages known as kephir and koumiss. Kephir is prepared from

[1] The ginger-beer plant is also known as Californian bees and by other popular names; both lately and after the Crimean War a tradition arose that it had been brought to this country by soldiers from overseas. Cf. J. Ramsbottom, *Trans. Brit. Myc. Soc.* 1920, p. 86.

cows' milk in the Caucasus, and koumiss from mares' milk in South Western Siberia. In both these cases the formation of alcohol depends on the yeast, and that of lactic acid is due to bacterial activity.

A different type of association is found in the preparation of the Japanese rice wine or sakè. Here the starch of the rice is hydrolysed by *Eurotium Oryzae* and the resultant sugar is fermented by a yeast.

In other cases zymase is secreted by filamentous fungi such as *Rhizopus nigricans*, *Penicillium glaucum*, and *Eurotium nigrum*. *Mucor racemosus* and certain other species when cultivated in sugar solution form ovoid cells which multiply by budding and cause active fermentation.

**Soot Fungi.** The soot fungi (*Meliola*, *Capnodium*, etc.), like the wild yeasts, are epiphytic saprophytes; they occur on leaves frequented by green fly and obtain their food from the "honey-dew" excreted by these insects. Their mycelium grows rapidly and forms a sooty coating on the leaves of the host, but does not become thick enough to injure them by excluding light.

## SAPROPHYTISM: BIBLIOGRAPHY

1892 WARD, H. MARSHALL. The Ginger-Beer Plant and the Organisms composing it. Phil. Trans. clxxxiii, p. 125.

1895 KLÖCKER, A. and SCHIÖNNING, H. Experimentelle Untersuchungen über die vermeintliche Umbildung des *Aspergillus oryzae* in einem Saccharomyceten. Centralbl. für Bakt. Abt. ii; i, p. 777.

1896 KLÖCKER, A. and SCHIÖNNING, H. Experimentelle Untersuchungen über die vermeintliche Umbildung verschiedenen Schimmelpilze in Saccharomyceten. Centralbl. für Bakt. Abt. ii; ii, p. 185.

1897 WARD, H. MARSHALL. On the Biology of *Stereum hirsutum*. Phil. Trans. clxxxix, p. 123.

1898 WARD, H. MARSHALL. *Penicillium* as a Wood-destroying Fungus. Ann. Bot. xii, p. 565.

1899 BIFFEN, R. H. A Fat-destroying Fungus. Ann. Bot. xiii, p. 363.

1899 CZAPEK, F. Zur Biologie der holzbewohnenden Pilze. Ber. deut. bot. Ges. xvii, p. 166.

1901 BIFFEN, R. H. On the Biology of *Bulgaria polymorpha*, Wett. Ann. Bot. xv, p. 119.

1901 GREEN, J. REYNOLDS. The Soluble Ferments and Fermentation. Cambridge Univ. Press.

1901 MASSEE, G. and SALMON, E. S. Researches on Coprophilous Fungi. Ann. Bot. xv, p. 313.

1902 VON SCHRENK, H. Decay of Timber and the Means of Preventing it. U.S. Dept. of Agr. Bureau of Plant Industry. Bull. 14, New York.

1903 MOLLIARD, M. Sur une Condition qui favorise la Production des Périthèces chez les Ascobolus. Bull. Soc. Myc. Fr. xix, p. 150.

1905 BULLER, A. H. R. The Destruction of Wooden Paving Blocks by the Fungus *Lentinus lepideus*, Fr. Journ. Econ. Biol. i, p. 1.

1906 BULLER, A. H. R. *Polyporus squamosus* as a Timber destroying Fungus. Journ. Econ. Biol. i, p. 101.

1907 FRASER, H. C. I. On the Sexuality and Development of the Ascocarp in *Lachnea stercorea*, Pers. Ann. Bot. xxi, p. 349.

1907 WELSFORD, E. J. Fertilization in *Ascobolus furfuraceus*. New Phyt. vi, p. 136.

1909 BULLER, A. H. R. Researches on Fungi. Longmans, Green & Co., London.
1909 BULLER, A. H. R. The Destruction of Wood by Fungi. Sci. Prog. xi, p. 1.
1909 CUTTING, E. M. On the Sexuality and Development of the Ascocarp in *Ascophanus carneus*. Ann. Bot. xxiii, p. 399.
1912 DALE, E. On the Fungi of the Soil. Ann. Myc. x, p. 452.
1912 DODGE, B. O. Methods of Culture and the Morphology of the Archicarp in Certain Species of the Ascobolaceae. Bull. Torrey Bot. Club. xxxix, p. 139.
1913 GODDARD, H. N. Can Fungi living in Agricultural Soil Assimilate free Nitrogen? Bot. Gaz. lvi, p. 249.
1913 McBETH, I. G. and SCALES, F. M. Destruction of Cellulose by Bacteria and Fungi. U.S. Dept. of Agr. Bureau of Plant Industry. Bull. 266, New York.
1914 RAMLOW, G. Beiträge zur Entwicklungsgeschichte der Ascoboleen. Myc. Centralbl. v, p. 537.
1915 BADEN, M. L. Observations on the Germination of the Spores of *Coprinus sterquilinus*, Fr. Ann. Bot. xxix, p. 135.
1915 SCALES, F. M. Some Filamentous Fungi tested for Cellulose destroying Power. Bot. Gaz. lx, p. 149.
1920 HALL, A. D. The Soil. John Murray, London, 3rd ed. (Fairy Rings, p. 278).

## PARASITISM

**Facultative Parasites.** Several fungi which are capable of passing through their whole development as saprophytes are also occasionally found on living plants as facultative parasites or hemi-saprophytes. It was first shown by de Bary that such fungi possess the power of disintegrating and killing the tissues in advance, so that they are not parasitic in any strict sense, but first kill the cells of their host and then live saprophytically upon the dead remains. This is well seen in *Botrytis cinerea*, the detailed knowledge of which is due to Blackman and Welsford, and to Brown. When the spores of *Botrytis cinerea* are placed in a drop of nutrient fluid on the leaves of the broad bean (*Vicia faba*), they show the first signs of a germ-tube in 2–3 hours; the outer walls of the developing tube soon become modified to form a mucilaginous sheath by means of which the hypha adheres to its substratum. After growing for a while along the surface of the leaf the germ-tube turns down and its tip, filled with dense protoplasm, becomes pressed against the cuticle where it may or may not swell somewhat and become spread out to form an enlargement or simple **adpressorium**; as growth continues the germ-tube is held firmly in place by its mucilaginous coat and the cuticle is ruptured mechanically by the pressure of its tip. The fungus now penetrates directly into an epidermal cell, or grows more or less horizontally in the subcuticular layers; in either case these layers become swollen, and in doing so appear to stretch the cuticle and make its penetration by other germ-tubes easier. As the hyphae make their way through the epidermis the cells of the palisade parenchyma become affected; their nuclei begin to disintegrate, the chloroplasts swell, and the starch almost disappears. In the bean a dark coloration is one of the characteristic signs of death, and this colour change spreads through the mesophyll in advance of the fungal hyphae. The hyphae, in fact, have been shown to secrete an enzyme which both disintegrates the walls of the host cells and causes the death of the protoplasts; it is however quite unable to affect the cuticle, the penetration of the outer walls of the epidermis being mechanical; a corresponding mechanism has been demonstrated in *Colletotrichum Lindemuthianum* and it may be inferred that the same occurs in other less fully investigated cases.

Penetration of the cuticle is, however, by no means a necessary preliminary to parasitism, whether obligate or facultative, for the hyphae of many fungi enter the host through the stomata, and in others, the so-called **wound parasites**, infection only takes place where a previous injury has exposed the internal tissues.

It is a curious point, and one deserving further investigation, that while the germ-tubes from certain fungal spores always make use of the stomata as a means of entrance, certain others completely ignore them, even though the germinating spore happens to lie close to a stoma; it is possible that in the former case a hydrotropic or negatively phototropic stimulus and in the latter a contact stimulus alone or in addition is operative. In *Botrytis cinerea* hyphae sometimes grow into an already seriously infected leaf by way of the stomata; this may be related to the fact that the stomatal space becomes charged with fluid due to the breaking down of the host cells.

A positively chemotropic reaction has often been suggested (Miyoshi) as explaining the entrance of the germ-tube into the leaf, but recent investigation (Fulton, Robinson, Graves) show that positively chemotropic sensitiveness is weak or absent in all the hyphae studied.

In *Botrytis* the mycelium commonly enters the cells of the host, but in other fungi it may develop wholly in the intercellular spaces, killing those cells with which it comes in contact and benefiting by the food materials that diffuse out through the dead protoplast; such forms may be described as saprophytic in the same sense as *Botrytis*.

As might be expected fungi whose attack is mainly directed to the elements of the wood flourish equally well on living and on dead tissue; the harm that they do to their host largely depends on the fact that they cut off the water supply to the regions beyond the infected area. This is the case with *Nectria cinnabarina*, and the fact that in this species the ascocarps are produced on dead tissue emphasizes its saprophytic or hemi-saprophytic character. We have here a case where the survival of the host is of no advantage to the attacking fungus.

**Obligate Parasites.** In the case of obligate parasitism, on the other hand, it is evident that the death of the host involves the death of the fungus, and it is to the interest of the parasite that death should be postponed, at least until the latter has itself made provision for reproduction.

The relations of the parasitic fungus to its host are extremely various; some, especially certain unicellular parasites (*Olpidium*, *Woronina*) go through their whole development inside the cells of the host, some live wholly in the intercellular spaces (*Gnomonia*) or under the cuticle (*Exoascus*, *Taphrina*) and obtain nutriment by osmosis; others possess an intercellular mycelium from which short branches are sent into the living cells of the host; these branches may become specialized to form **haustoria** of limited growth and more or less definite form (*Cystopus*, *Ustilago*, etc.). In all the above cases the parasite is inside the host and may be described as **endophytic**.

In a certain number of forms the development of the parasite is external, and may be described as **epiphytic**. This is the case in many mildews (Erysiphaceae) which obtain their food supply by sending haustoria into the epidermal cells of the host; it is also the case in the Laboulbeniales where food is absorbed in many species through the unbroken membranes of the host, and where the parasite probably causes a minimum of damage and inconvenience.

The influence on the parasite of its method of life is already shown at a very early stage in development. Thus the conidia or zoosporangia of *Cystopus candidus* germinate better at low temperatures than high; their minimum is near zero, their maximum about 25° C. and their most favourable temperature, under ordinary circumstances, 10° C. The spread of the fungus by zoospores depends on the presence of water on the foliage of the host, a fall of temperature leads to the deposition of dew, thus providing the condition for the activity of the zoospores, and at the same time serves as a stimulus to their development.

In contrast to the above the uredospores of *Puccinia dispersa*, which give rise to a germ-tube directly, germinate between 10° and 27° C. but most readily at about 20° C.; the high temperatures appropriate to the germination of the spores of coprophilous fungi may also be recalled in this connection.

The relation between the parasite and its host may be strictly localized, as in the Laboulbeniales and in those Archimycetes which enter a single cell and complete their development within it; or the parasite may spread far beyond the point of infection, extending through or over the existing parts of the host and keeping pace with its growth when the new tissues are developed; in this way the mycelium of certain of the Ustilaginales is found year after year in the tissues of the herbaceous host, dying down when it prepares for the winter, and growing with its growth; perennial mycelia are not uncommon in Exoascaceae, Uredinales and other forms which infect trees.

The parasite may modify the host tissues by its invasion, chiefly in the direction of abnormal growth or **hypertrophy**. The simplest instance of such an effect is the enlargement of a single cell due to the entrance of the parasite; this is found in the infection of various algae and fungi by *Olpidium*, and of the dandelion and other angiosperms by *Synchytrium*; in the latter instance the cells surrounding the seat of infection are also enlarged.

In many cases fresh cell formation as well as enlargement of the host cells takes place; this may be limited to the neighbourhood of the infected spot so that the host organ becomes locally deformed; thus "peach leaf curl" and many similar abnormalities are formed by *Exoascus* and its allies; and irregular rose-coloured blisters are produced on the leaves of Ericaceae infected by *Exobasidium*. More elaborate deformations are produced by some of the Ustilaginales, and, in the case of "witches'-brooms," by the rusts and Exoascaceae.

A witch's-broom is a bunch of modified twigs, caused usually by insects, but sometimes by fungi. In the latter case it is the product of a lateral bud, which, stimulated by the presence of the fungus, or by the food material which the cells of the fungus deflect from its proper course, grows out to form an abnormally dense bush of twigs; its leaves are produced earlier than those of healthy branches and, even in the case of normally evergreen conifers, are deciduous, falling off at the end of each season. Here the specialized shoot, in spite of its contained parasite, appears to flourish, though to the detriment of the rest of the tree; it may indeed be suggested that something approaching symbiosis has been established, but in this relationship the fungus is clearly the dominant partner[1].

## SYMBIOSIS

The physiological conditions under which the thallus of a lichen is built up are somewhat similar; the algal cells appear healthy and are capable of vegetative multiplication, but the fact that the fungus alone is concerned in the development of the fructification sufficiently indicates its supremacy. In the case of *Thermutis orlutina* the alga is devoured by the fungus in preparation for the formation of the fruit.

In **mycorhiza**, that is in the structures formed by the association between the mycelium of a fungus and the roots (or other organs) of one of the higher plants, the advantages of the symbiotic relation often belong less to the fungus than to its partner so that the vascular plant may actually become dependent on its fungal associate and unable to develop in its absence. The mycelium may be either **endotrophic**, forming a skein of branched and interwoven filaments in the cells of the host and sending comparatively few hyphae to the exterior; or it may be **exotrophic**, that is to say mainly external in development.

**Endotrophic Mycorhiza**. An extreme case of obligate symbiosis has been described by Rayner for the ling, *Calluna vulgaris*, which grows in association with a fungus resembling the members of the genus *Phoma* in

[1] For bibliography, see p. 19.

its morphological characters. The mycelium not only extends through the roots and colourless parts, but grows into the subaerial tissues of the stem, leaves, flowers and fruit. Moreover the seedling is regularly infected on germination by the hyphae which have penetrated the seed coat. When deprived of its fungus by sterilization of the seed, the seedling fails to develop, its root system in particular being inhibited; when subsequently infected from artificial culture it renews its growth. If, however, a weak seedling is inoculated from a vigorous culture it is completely parasitized and destroyed; the fungus in such a case may be regarded as having escaped from the control of its partner.

The fungi present in the orchids investigated by Bernard were referred to the form genus *Rhizoctonia*[1], certain species of which have been shown to be conidial stages of the basidiomycete *Hypochnus*. It is significant that conidia are never produced on a healthy host plant, though they can be obtained when the mycelium is grown in culture; in the cells of the host a tangled weft of hyphae develops and ultimately undergoes digestion, forming an amorphous mass. In *Odontoglossum* the mycelium does not enter the stem, and in *Vanda* and its allies it is confined to the perennial roots; in the Cattleyeae the roots are deciduous and their disappearance is followed by an autonomous phase; in *Bletilla hyacinthina*, also, the mycelium never invades the green, superficial rhizome, but the young roots are regularly infected as they reach the length of a few centimetres. In winter the orchid is represented by its rhizome alone and its activity at the beginning of fresh growth is consequently autonomous; the symbiotic phase follows on the development of the new roots and lasts about six months, covering the period of the maturation of the seeds. During the autonomous phase the fungus vegetates in the soil and loses, to some extent, its "virulence" or power of infection.

*Bletilla* differs from the cases previously described in that its symbiosis is facultative and development can take place without infection, but the seedlings grow slowly and are delicate; inoculation by a mycelium greatly reduced in virulence has little effect, the hyphae enter a few cells and are at once digested; a mycelium, on the other hand, which has attained a high degree of virulence, penetrates at once to the region of attachment of the suspensor and instead of undergoing premature digestion spreads through the host cells while the seedling grows rapidly and its lower part swells to form a protocorm.

The restriction of the mycelium to the non-chlorophyllous regions of these orchids is not accidental, for the contents of certain of the stem cells have a poisonous effect on the fungus. It follows that here the mycelium cannot penetrate into the ovary and the infection of the germinating seed is consequently a matter of chance; hence the difficulty frequently experienced in inducing the germination of orchids.

[1] Burgeff, in 1909, proposed the new generic name *Orcheomyces* for these fungi.

Kusano has shown that the rootless, saprophytic orchid, *Gastrodea elata*, becomes dependent on the formation of mycorhiza only on the incidence of its flowering period; it is capable of vegetative growth without infection, and the distribution of the fungus in its tissues is strictly limited; the fungus in this case is *Armillaria mellea*, a well-known facultative parasite.

This may be regarded as a relatively early stage in the development of endotrophic symbiosis in which the angiosperm succeeds in utilizing as a factor in its own nutrition the mycelium of the attacking parasite. It suggests that in this relationship we are dealing not with an association of two organisms for the benefit of both, but with a struggle between the would-be parasite and the host which controls, makes use of, and finally becomes dependent upon it. The balance of power is often very delicate, and any weakening on the part of the angiosperm gives the fungus an opportunity to assume the parasitic habit; thus the endotrophic fungus regularly becomes parasitic on the old stamens and corolla of *Arbutus* and on the dying leaves. Endotrophic mycorhiza occurs in the Ericaceae and in their allies showing, so far as investigation has gone, the same extreme conditions as in *Calluna*; in certain Gentianaceae, Solanaceae, Labiatae, Umbelliferae, Ranunculaceae and Liliaceae, in most if not all orchids, in species of *Viola* and *Arum*, and, according to the researches of von Tubeuf, in gymnospermous trees other than the Abietineae.

Endotrophic mycorhiza is also found in certain liverworts and mosses, and in many of the Pteridophyta. It is well developed in the prothallus of *Lycopodium* and in the leafy plant of species of both *Lycopodium* and *Selaginella*, in the roots of *Cyathea* and of several of the Marattiaceae, and in both the prothallus and sporophyte of the Psilotaceae and Ophioglossaceae. In most of these cases it does not appear to be essential to the nutrition of the sporophyte since the latter seldom shows a correlated reduction of the assimilatory apparatus, but it is often a principal factor in the nutrition of the prothallus, which in the presence of mycorhiza may be subterranean and lacking in chlorophyll. Mycorhiza has been recorded in a number of fossil plants.

Bernard has suggested a relationship between tuberization and the presence of an endotrophic mycelium. It is significant that very many symbiotic plants, including notably the orchids and the sporophyte and gametophyte of the lycopods, tend to assume a tuberous habit or to develop bulbs or protocorms. Bernard's researches have disclosed mycorhiza in tuber-forming species of *Solanum*, indicating a similar origin for the potato.

**Exotrophic Mycorhiza.** The majority of our forest trees, including Abietineae, Salicaceae, Fagaceae and Betulaceae, as well as some other plants, possess an exotrophic mycelium forming a dense felt over the apical parts of the infected roots; under this influence the development of root hairs is impeded and characteristic short, coral-like branches are formed. *Sarcodes sanguinea* and *Monotropa Hypopitys* are non-green holosaprophytes with exotrophic mycorhiza; in the former the whole root system is covered by the mycelium of the fungus, in the latter the apices alone are free.

In other cases infection seldom takes place in all roots; it is extensive in soils rich in vegetable debris, and is absent in poor soils free from humus; under these circumstances root hairs are formed and the roots function in the normal way. When present it would appear that the mycorhiza not only absorbs water and dissolved salts, but the vascular plant is further enabled by its means to utilize directly the organic remains in the soil.

The relation between an exotrophic fungus and its host, at least when the latter contains chlorophyll, would appear to be much more casual than is the case with endotrophic forms; like the latter it may arise as an attempted parasitism on the part of the fungus, controlled and utilized by the vascular plant. The absence of exotrophic mycorhiza in poor soils may depend at least as much on the absence of saprophytic mycelia capable of causing infection, as on the fact that a fungal associate would, under the circumstances, be of little value to the green plant. Exotrophism would seem to be more to the advantage of the fungal partner than endotrophism, since fructifications are often found on mycelia in exotrophic association with the roots of vascular plants; and some subterranean species, such as the truffles, fruit only in the neighbourhood of appropriate trees.

The fungi concerned in these curious relationships include representatives of all the great groups; the endotrophic mycelium of the prothalli of *Lycopodium* has been referred to the genus *Pythium*, and that of the Marattiaceae to *Stigeosporium*, a genus nearly related to *Phytophthora*, that of several orchids to *Rhizoctonia* (= *Hypochnus*?) and of others to *Nectria*; the mycelium of species of *Elaphomyces* forms mycorhiza with the roots of *Pinus* and other conifers; that of the *Boleti* with conifers and grasses and with willow, poplar, hornbeam and birch, that of *Tricholoma terreum* with beeches and firs, that of *Lactarius piperatus* and of species of *Cortinarius* with beeches and oaks, that of species of *Geaster* with conifers. Reference has already been made to the association of *Armillaria mellea* with the orchid *Gastrodea elata*, and of *Phoma*-like species with the Ericaceae.

Finally there is little doubt that several of the Tuberaceae and Hymenogasteraceae are frequent constituents of mycorhiza.

## PARASITISM AND SYMBIOSIS: BIBLIOGRAPHY

1863 DE BARY, A. Recherches sur le développement de quelques champignons parasites. Ann. Sci. Nat. xx, p. 95.

1888 WARD, H. MARSHALL. Some Recent Publications bearing on the Sources of Nitrogen in Plants. Ann. Bot. i, p. 325.

1890 OLIVER, F. W. On *Sarcodes sanguinea*, Torr. Ann. Bot. iv, p. 303.

1891 FRANK, B. Ueber die auf Verdauung von Pilzen abzielende Symbiose der mit endotrophen Mycorhizen begabten Pflanzen, so wie der Leguminosen und Erlen. Ber. d. deutsch. Bot. Ges. ix, p. 244.
1894 MIYOSHI, M. Ueber Chemotropismus der Pilze. Bot. Zeit. lii, p. 1.
1902 LANG, W. H. On the Prothalli of *Ophioglossum pendulum* and *Helminthostachys zeylanica*. Ann. Bot. xvi, pp. 28 and 36.
1904 CAVERS, F. On the Structure and Biology of *Fegatella conica*. Ann. Bot. xviii, p. 95
1904 MASSEE, G. On the Origin of Parasitism in Fungi. Phil. Trans. B. cxcvii, p. 7.
1904 WEISS, F. E. A Mycorhiza from the Lower Coal Measures. Ann. Bot. xviii, p. 255.
1905 GALLAUD, I. Études sur les mycorhizes endotrophes. Rev. Gén. Bot. xvii, p. 1.
1906 FULTON, H. R. Chemotropism of Fungi. Bot. Gaz. xli, p. 81.
1908 BOWER, F. O. The Origin of a Land Flora. Macmillan & Co., London, pp. 240, 477 *et seq.*
1909 BERNARD, N. L'évolution dans la symbiose des Orchidées et leurs champignons commensaux. Ann. Sci. Nat. 9me sér. ix, p. 1.
1909 BURGEFF H. Die Wurzelpilze der Orchideen. Fischer, Jena.
1911 BERNARD, N. Sur la fonction fungicide des bulbes d'Ophrydées. Ann. Sci. Nat. 9me sér. xiv, p. 221.
1911 BERNARD, N. Les mycorhizes des *Solanum*. Ann. Sci. Nat. 9me sér. xiv, p. 235.
1911 KUSANO, S. *Gastrodea elata* and its Symbiotic Association with *Armillaria mellea*. Journ. Coll. Agric. Tokio, iv, p. 1.
1911 MELHUS, I. E. Experiments on Spore Germination and Infection in Certain Species of Oomycetes. Univ. Wisc. Agric. Expt. Sta. Research Bull. xv, p. 25.
1914 ROBINSON, W. Some Experiments on the Effect of External Stimuli on the Sporidia of *Puccinia malvacearum* (Mont.). Ann. Bot. xxviii, p. 331.
1915 RAYNER, M. C. Obligate Symbiosis in *Calluna*. Ann. Bot. xxix, p. 97.
1915 BROWN, W. Studies in the Physiology of Parasitism. I. Ann. Bot. xxix, p. 313.
1916 BLACKMAN, V. H. and WELSFORD, E. J. Studies in the Physiology of Parasitism. II. Ann. Bot. xxx, p. 389.
1916 BROWN, W. Studies in the Physiology of Parasitism. III. Ann. Bot. xxx, p. 399.
1916 GRAVES, A. H. Chemotropism in *Rhizopus nigricans*. Bot. Gaz. lxii, p. 337.
1916 RAYNER, M. C. Recent Developments in the Study of Endotrophic Mycorhiza. New Phyt. xv, p. 161.
1917 BROWN, W. On the Physiology of Parasitism. New Phyt. xvi, p. 109.
1917 DUFRENOY, J. The Endotrophic Mycorhiza of Ericaceae. New Phyt. xvi, p. 222.
1917 WEST, C. On *Stigeosporium Marattiacearum* and the Mycorrhiza of the Marattiaceae. Ann. Bot. xxxi, p. 77.
1919 DEY, P. K. Studies in the Physiology of Parasitism. V. Infection by *Colletotrichum Lindemuthianum*. Ann. Bot. xxxiii, p. 305.
1921 RAMSBOTTOM, J. Orchid Mycorhiza. Catalogue of J. Charlesworth & Co., Haywards Heath, p. 1.

## SPECIALIZATION OF SAPROPHYTISM AND PARASITISM

Fungi vary very much in the extent to which they are adapted or restricted to a particular habitat. In some species the range is very wide, as in the case of *Eurotium herbariorum* or *Penicillium glaucum* which may occur under suitable conditions of temperature and moisture on almost any plant remains, on plant products such as bread or jam, or on substances of animal origin and, in the case of *Penicillium*, especially on cheese.

*Eurotium herbariorum* and some other species are even found in the human ear, where they produce the condition known as *otomycosis aspergillana*, but they develop on the secretions and not in the living cells, so that, although they give rise to a disease, they are not true parasites.

Similarly *Synchytrium aureum* infects all sorts of dicotyledons and *Phyllactinia Corylea* occurs on the leaves of many trees.

In other cases the range is somewhat narrowed; many species of *Hydnum* are found only in fir woods, *Pyronema confluens* and a number of other fungi occur in nature only on burnt ground, *Onygena equina* is restricted to the hoofs and horns of various animals, and several species (*Sordaria macrospora*, *Podospora coprophila*, *Ascobolus immersus*, *Ascophanus equinus*, etc.) develop upon different kinds of dung but on no other substratum. Some of the coprophilous fungi on the other hand may appear in addition on other substances, especially on those rich in cellulose—*Ascophanus carneus* has been recorded upon paper or rope, *Gymnoascus Reesii* on wasps' nests, *Xylaria Tulasnei* on soil.

Parasites again may be limited to hosts of a particular family; *Cystopus candidus* to the Cruciferae, *Claviceps purpurea* to grasses, *Ustilago violacea* to the Caryophyllaceae.

Still greater restriction is observed in the case of such saprophytes as *Erinella apala* on the dead stems of species of *Juncus* and *Pilacre faginea* on rotten beech wood, and in the case of the parasites which attack the species of a single genus, such as *Rhytisma Acerinum* on *Acer*, *Polystigma rubrum* on *Prunus*, *Peronospora Euphorbiae* on *Euphorbia*.

Even more definite is the specialization of fungi which are capable of obtaining nutriment only from a single species; this occasionally happens among saprophytes, such as *Dasyscypha clandestina* on dry stems of *Rubus idaeus*; it may be surmised that in many such cases new habitats will ultimately be brought to light.

Among parasites restriction to a single species is very common; *Sclerotinia tuberosa* forms its sclerotium attached to the rhizome of *Anemone nemorosa*, *Exoascus Betulae* develops on *Betula alba*, and *Ustilago Maydis* on *Zea Mays*.

De Bary gave to this state the name **monoxeny**, in contradistinction to the term **polyxeny**, which he applied to cases where hosts of several different species may be attacked. Every sort of intermediate grade may exist between those outlined above; parasites may attack only two or three closely related members of a genus (*Exobasidium Rhododendri* on *Rhododendron hirsutum* and *R. ferrugineum*), they may attack a genus and one or two species belonging to related genera, they may be capable of development on certain genera of a family and not on the rest, or again, while commonly restricted to a particular family, they may occur also on members of neighbouring groups; thus *Phytophthora infestans* is found usually on the Solanaceae and exceptionally on the scrophulariaceous species *Anthocercis viscosa* and *Schizanthus Grahami*.

In all the cases of parasitism considered above, the fungus, whether monoxenous or polyxenous, is capable of passing through its whole existence on a single host. A different type of specialization appears in those fungi which require two host species for the completion of their life-history, and produce characteristic spores on each. Such forms are said to be **heteroecious** in contradistinction to the **autoecious** species on a single host.

**Heteroecism.** Heteroecious species may be either monoxenous or polyxenous, that is to say they may be limited in relation to either or both their stages to a single species of host, or they may be capable of occurring on a number of different forms. Heteroecism is known in *Sclerotinia Ledi* but in no other fungi except the Uredinales or rusts, and its full consideration must be postponed till that group has been described. It may be considered here as one of the possible alternatives confronting a parasite on a host which dies down early in the year. Under such circumstances the parasite may continue its development on the dead tissue, it may await in the form of resting spores the reappearance of its host, or its spores may prove capable of germinating on a new host species, and it may carry on its development as a parasite and reinfect its original host next spring, thus becoming heteroecious. It must be noticed, however, that by no means all the existing spring or summer hosts of heteroecious fungi die down early in the year, so that other and possibly secondary factors will have to be considered.

**Biological species.** There is no doubt that within wide or narrow limits related host plants are apt to show common susceptibility to infection; this is well exemplified by *Puccinia Malvacearum* which was first observed in this country on cultivated hollyhocks in 1873 and has since established itself on the indigenous species of *Malva*, *Althea* and *Lavatera* as well as in greenhouses on *Abutilon*.

Common susceptibility may even be used as a criterion of relationship, so that liability to the attacks of *Piptocephalis Freseniana*, the obligate parasite of the Mucorales, has afforded a means of distinguishing new members of that group.

These facts point to a definite adaptation on the part of the fungus to its habitat; this adaptation may be very simple, a species, for example, would not be likely to occur on dung unless its spores could pass uninjured through the alimentary canal of an animal, or it may reach the complexity of a delicately balanced reaction between host and fungus, as in some of the mycorhiza described in the preceding pages.

Specialization has been most fully investigated and has possibly reached its highest levels in the adaptation of various rusts and mildews to their graminaceous hosts, but it is shown in varying degrees by other fungi[1]. Like

[1] Wormald has shown that there are two biological species of *Monilia* (*Sclerotinia*) *cinerea*, one of which produces a Blossom Wilt and Canker Disease on the apple tree while the other is unable

many fruitful mycological discoveries this was foreshadowed by de Bary, who in 1863, noticed that the structural differences between the aecidia of *Chrysomyxa Rhododendri* and *C. Ledi* were so slight that he regarded these as "rather biological than morphological species."

Thirty years later Eriksson recognized that the rust of wheat, *Puccinia Graminis*, which infects wheat, barley, rye, oats and various wild grasses, is a collective species, consisting of a number of biological forms which, though they differ in no recognizable structural character, yet differ in the powers of infection of their spores, since uredospores grown on wheat are incapable of directly infecting rye, barley or oats, those on oats cannot directly infect wheat, rye or barley, and those on barley and rye, though they can infect both rye and barley, will not develop if sown on oats or wheat. Now *Puccinia Graminis* is a heteroecious species producing two kinds of spores (uredospores and teleutospores) on grasses, and aecidiospores in cluster-cups on the barberry. But, though all the different biological forms alike develop their cluster-cups on the barberrry, Eriksson found that they remained constitutionally distinct, for aecidiospores derived from the form upon oats proved capable of infecting among cereals only oats, aecidiospores from the form upon rye or barley, only rye or barley and so on.

In other words each form of *Puccinia Graminis* is so closely adapted to the particular cereal on which it occurs that its spores can only attack successfully and directly that particular graminaceous host or a limited number of its immediate allies.

Marshall Ward, working with the uredospores of *Puccinia dispersa*, made clear that the susceptibility or immunity of the host does not depend on structural characters, and suggested rather the existence of enzymes or toxins or both in the cells of the fungus, and of antitoxins or similar substances in the cells of the host.

This hypothesis has been greatly strengthened by the work of Marchal and of Salmon on the Erysiphaceae or mildews in which group both the conidia and ascospores of biological species are similarly specialized in their powers of infection. Not only are there no structural peculiarities in the resistant hosts of these fungi, but Salmon was able, by suitable treatment, to break down their resistance. This may be achieved in various ways: (1) a minute piece of tissue, including the epidermis and the greater part of the mesophyll, is cut with a razor from one side of the leaf and spores are sown on the opposite side; (2) the leaf to be inoculated is touched for a few seconds on the upper surface with a red-hot knife and the spores are sown on the lower surface opposite the burnt spot.

The result of such treatment is the ready infection of host species

to cause infection except of a flower directly inoculated. Biological species have also been identified among smuts, and by Diedicke in *Pleospora*.

normally immune from the attacks of the biological form used. Thus if conidia of *Erysiphe Graminis* growing on wheat, are sown on uninjured leaves of wheat and barley, the result is the infection of the wheat but never of the barley; yet the conidia grown on wheat readily infect the uninjured surface of a cut or burnt barley leaf; in the case of the burnt leaf, where the whole thickness of the leaf in the burnt region becomes discoloured and apparently dead, the mycelium is found on the living cells which border the altered patch. Since the cuticle, hairs and other anatomical characters of the epidermis on which the spores are sown are not affected by the treatment of the cut leaf, it is clear that resistance does not depend on such factors; it must be referred, as in the case of rusts, to the physiological condition of the cells or of their contents. In nature injuries caused by insects are sufficient to destroy in the same way the resistance of the potential host.

These facts have a practical bearing since diseases on the weed grasses surrounding a field of corn, even if they are not able directly to affect the uninjured leaves of the crop, may establish themselves on injured tissues.

The various hosts of a given morphological species of fungus differ very much in their susceptibility to infection by the different biological forms of the parasite; thus *Bromus racemosus*, though markedly susceptible to its own form of *Erysiphe Graminis*, is completely immune against infection by conidia of the same species grown on *B. commutatus*, *B. interruptus*, *B. velutinus* and others. This immunity is particularly remarkable in the case of

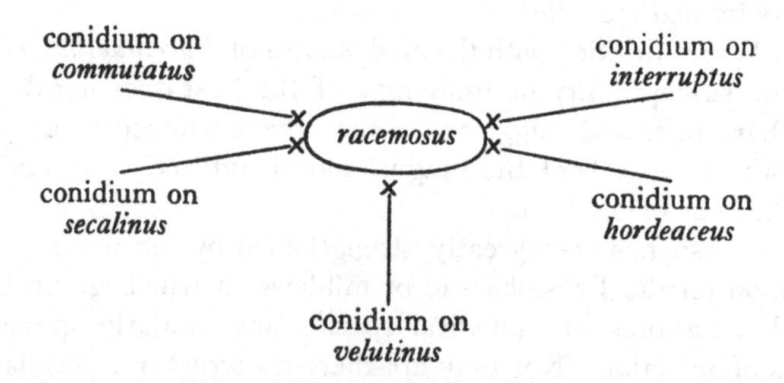


the conidia from *B. commutatus*, since *B. racemosus* is morphologically so close to *B. commutatus* that it is regarded by most systematists as no more than a variety of that species.

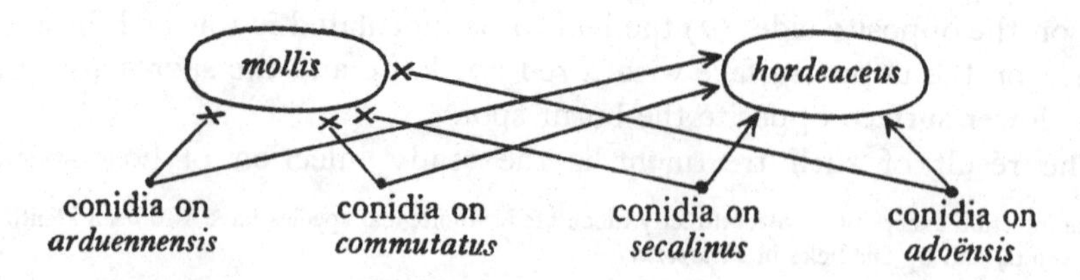

Another remarkably resistant species is *Bromus mollis*, yet the so-called *B.* "*hordeaceus*," the seed of which was sent to Cambridge from Petrograd, and which is morphologically indistinguishable from *B. mollis*, is nevertheless susceptible to infection by the biological species which *B. mollis* is able to resist. In other words the morphological species *B. mollis* includes two groups or races possessing distinct physiological (or constitutional) characters and respectively immune and susceptible to infection.

In the case of certain parasitic fungi, and especially of *Puccinia glumarum*, the yellow rust on wheat, Biffen has shown that resistance to the attacks of the parasite is a recessive character in the Mendelian sense. When a variety susceptible to rust was crossed by another practically immune from it, the offspring was at least as much infected as the original rusty type. In the next generation segregration took place in the ordinary way, three-quarters of the plants being rusty and the remaining quarter standing green and uninjured among them. In relation to this investigation Marryat showed that the rust hyphae are checked after entering the stomata of the resistant plants either by the death of the host tissue locally, accompanied by the starvation and death of the parasite, or, after a more protracted struggle, by the gradual degeneration of the invading hyphae. If, as has been suggested above, resistance depends on the presence of an antitoxin, the dominance of susceptibility in this case must be taken to indicate that the development of the antitoxin is inhibited by the presence of some additional factor in the dominant forms. Cases may be expected where, in the absence of an inhibitor, resistance is dominant and depends on the presence of the antitoxin or its progenitor. Such a case is suggested by the work of Biffen on the inheritance of immunity to ergot, but here two factors appear to be involved.

The susceptible races have in some instances an importance beyond that implied in their own liability to infection; it has been suggested that they may serve as a bridge by which the fungus can pass from its original host to a species resistant to direct attack.

This was indicated by the work of Marshall Ward on *Puccinia dispersa*, and by Salmon using *Bromus* "*hordeaceus*" as a bridging host. Inoculation experiments showed that the form of *Erysiphe Graminis* on *Bromus racemosus* was incapable of developing on uninjured *B. commutatus* but that it never

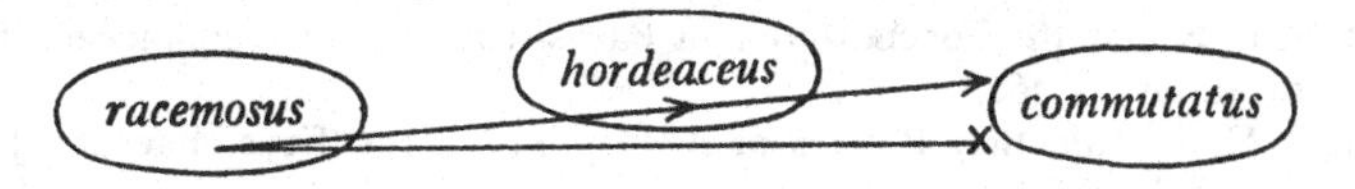


failed to produce full infection on *B.* "*hordeaceus*." The mycelium on its new host gave rise in due course to conidia and some of these, when transferred to *B. commutatus*, succeeded in establishing themselves and produced full infection in eight days.

Freeman and Johnson, in 1911, described barley as a bridging host for biological forms of *Puccinia Graminis* on other cereals and, in the same year, Pole Evans found that the heterozygote between strains immune and susceptible to infection by black rust was even more susceptible than the susceptible parent and was capable of acting as a bridging species from which the immune parent could be infected.

On the other hand an increasing number of investigators, working with isolated strains under rigidly controlled conditions, have failed to confirm the existence of bridging species and, according to Stakman and his assistants, there is no evidence that biological species are changed by their sojourn on a particular host. Such a change would imply a certain physiological plasticity on the part of the fungus, since it would be capable of becoming acclimatized, on the bridging host, to conditions similar to those awaiting it on the form resistant to direct attack.

In the case of *Puccinia Graminis* the problem is further complicated by the fact that a considerable number of biological forms are parasitic on the varieties of wheat, so that the same variety may be susceptible in one locality and immune in another according to the distribution of the parasite.

## SPECIALIZATION OF PARASITISM ETC.: BIBLIOGRAPHY

1863 DE BARY, A. Recherches sur le développement de quelques champignons parasites. Ann. Sci. Nat. sér. 4, xx, p. 5.

1894 ERIKSSON, J. Ueber die Specialisirung des Parasitismus bei den Getreiderostpilzen. Ber. der deut. Bot. Ges. xii, p. 292.

1898 ERIKSSON, J. A General Review of the Principal Results of Swedish Research into Grain Rust. Bot. Gaz. xxv, p. 26.

1902 DIEDICKE, H. Ueber den Zusammenhang zwischen *Pleospora*- und *Helminthosporium*-Arten. Centralbl. f. Bakt. ix, Abt. ii, p. 317.

1902 WARD, H. MARSHALL. On the Relations between Host and Parasite in the Bromes and their Brown Rust, *Puccinia dispersa* (Erikss.). Ann. Bot. xvi, p. 233.

1902 WARD, H. MARSHALL. Experiments on the Effect of Mineral Starvation on the Parasitism of the Uredine Fungus, *Puccinia dispersa*, on Species of *Bromus*. Proc. Roy. Soc. lxxi, p. 138.

1902–3 MARCHAL, E. De la Spécialisation du parasitisme chez l'*Erysiphe Graminis*. Comptes Rendus, cxxxv, p. 210, and cxxxvi, p. 1280.

1903 WARD, H. MARSHALL. Further Observations on the Brown Rust of the Bromes, *Puccinia dispersa* (Erikss.) and its adaptive Parasitism. Ann. Myc. i, p. 132.

1903 SALMON, E. S. On the Specialization of Parasitism in the Erysiphaceae. Beihefte 2, Bot. Central. xiv, p. 261.

1903 SALMON, E. S. Infection Powers of Ascospores in the Erysiphaceae. Journ. Bot. xli, pp. 159 and 204.

1904 MASSEE, G. On the Origin of Parasitism in Fungi. Phil. Trans. B. cxcvii, p. 7.

1904 SALMON, E. S. Cultural Experiments with Biologic Forms of the Erysiphaceae. Phil. Trans. B. cxcvii, 229, p. 107.

1904 SALMON, E. S. On *Erysiphe Graminis* DC. and its adaptive Parasitism within the Genus *Bromus*. Ann. Myc. ii, p. 255.

1905 WARD, H. MARSHALL. Recent Researches on the Parasitism of Fungi. Ann. Bot. xix, p. 1.
1907 BIFFEN, R. H. Studies in the Inheritance of Disease Resistance. I. Journ. Agr. Sci. ii, p. 109.
1907 MARRYAT, D. C. E. Notes on the Infection and Histology of two Wheats immune to the Attacks of *Puccinia glumarum*, yellow Rust. Journ. Ag. Sci. ii, p. 129.
1911 FREEMAN, E. M. and JOHNSON, E. C. The Rusts of Grains in the United States. U.S. Dept. Ag. Bureau of Plant Industry. Bull. 216.
1911 POLE EVANS, I. B. South African Cereal Rusts, with Observations on the Problem of Breeding Rust-resisting Wheats. Journ. Ag. Sci. iv, p. 95.
1912 BIFFEN, R. H. Studies in the Inheritance of Disease Resistance. II. Journ. Ag. Sci. iv, p. 421.
1914 VARILOV, N. I. Immunity to Fungous Diseases as a Physiological Test in Genetics and Systematics, exemplified in Cereals. Journ. Genetics, iv, p. 49.
1918 STAKMAN, E. C., PIEMEISEL, F. J., and LEVINE, M. N. Plasticity of Biologic Forms of *Puccinia graminis*. Journ. Ag. Research, xv, p. 221.
1919 WORMALD, H. The 'Brown Rot' Disease of Fruit Trees, with Special Reference to two Biologic Forms of *Monilia cinerea*, Bon. I. Ann. Bot. xxxiii, p. 361.

## REACTIONS TO STIMULI

Among the fungi, response takes place to a large number of external stimuli, most of which are concerned with nutrition and the distribution of the spores. A special series of reactions which demands further investigation takes place in relation to the formation, approach and fusion of the sexual organs. The stimulus in question may be effective very early in development for de Bary found that the presence of the oogonium in *Pythium de Baryanum* stimulates the formation of antheridia and Blakeslee observed directive growth or "zygotaxis" in morphologically undifferentiated hyphae of *Mucor Mucedo*. In *Mucor* and its allies the formation of the gametangia follows on the contact of appropriate branches. The stimulus inducing directive growth is presumably chemical, as in other and better known cases of the approach of gametes or gametangia; but we have at present no knowledge of the substances concerned or of much more than the fact that the reaction occurs. In some of the higher fungi, where the antheridium is liberated as a spermatium, it would appear to be carried passively to the female structure, but in *Zodiomyces* among the Laboulbeniales and perhaps in a few other Ascomycetes (*Ascobolus carbonarius*, *Lachnea cretea*) the trichogyne moves towards the male organ.

**Chemotropism.** The most marked chemotropic reaction of vegetative hyphae and germ-tubes is a negative one; they tend to grow away from the products of their own metabolism or so-called "staling" substances. Clark, in 1902, concluded that *Rhizopus nigricans*[1] is negatively chemotropic to some secretion of its own mycelium and that the negative response is much

[1] *Rhizopus nigricans*, Ehrenb. = *Mucor stolonifer*, Ehrenb.

greater than any positively chemotropic reaction towards food or oxygen. Fulton, in 1906, found for a number of different germ-tubes that they tend to turn from a region in which hyphae of the same kind are growing, to one destitute of hyphae or in which hyphae are less abundant. A very simple example of this reaction is found in the circular growth of mycelia both in nature and in artificial culture. In so far as a clear field is available, hyphae tend to grow equally in all directions from the point where infection took place. The same factor may account, as Stevens and Hall have suggested, for the alternate dense and sparse zones which characterize many fungal colonies and are independent of changes in light and temperature. Energetic growth results in the deposition of katabolic substances, and growth is correspondingly reduced till a few scattered hyphae pass beyond the inhibiting influence and give rise to a new ring of richly branched mycelium. In older colonies the germination of fresh spores outside the zone of staling substances doubtless adds to this effect.

The very definite character of the reaction is demonstrated by the experiments of Graves, who used the germ-tubes of *Rhizopus nigricans* on contrasted agar media separated by a perforated mica plate. Hyphae developing in agar made up with turnip juice grew towards the perforations which led to fresh but otherwise identical agar on the other side of the mica plate.

Hyphae under similar conditions turned from turnip juice, in which growth had taken place, to plain agar, or indeed to any fresh substance that was not in itself definitely repellant, even if, as in the case of plain agar, its nutritive value was low.

On the other hand, the hyphae turned away from the perforations in the mica plate if these led to nutritive agar on which mycelium had already developed and if the substratum on which the germ-tubes were growing was by comparison fresh. This was the case even after the mycelium on the staled agar had been taken away, the products of its metabolism which remained in the agar having themselves a repellant effect. This effect was removed by exposure to a temperature of 100° C., showing that the staling substances are altered or destroyed by heat.

A somewhat different example of the inhibiting effect of these katabolic staling substances was observed by Balls for the "sore-skin" fungus of cotton. When the nutritive medium is limited in extent, growth comes to an end more rapidly at high temperatures than at low. This is due not to the earlier exhaustion of the available food supply, but to the more rapid accumulation of staling substances when growth is accelerated. Media in which growth has come to an end prove capable of supporting further growth if diluted with an equal quantity of water. By this means the staling substances are sufficiently diluted to lose their inhibiting effect, while the nutritive materials, even though diluted, are still sufficient to support growth.

By the addition of staled media to his cultures Balls was able to show that accumulation of staling substances was a limiting factor in the growth of his fungus.

The products of past metabolism are not the only factors which exercise a negatively chemotropic influence; Miyoshi showed that hyphae tend to grow away from acids, alkalis, alcohol, toxic compounds and certain neutral salts.

He also brought forward evidence of the positively chemotropic effect of a number of salts and of sugar and other nutritive materials; but his results are vitiated by the fact that he was unaware of the importance of staling substances and accepted, as due to the attraction of the new medium, curvatures which really depended on the repellant effect of the old. Nevertheless, positive chemotropism exists and may be observed when the staling substances have been taken into account.

Thus Graves found that, while all the germ-tubes of *Rhizopus nigricans* turned from the turnip juice agar on which they had been growing to fresh turnip juice agar, only 60 to 90 °/₀ turned to fresh plain agar, and, when growth had just begun in the turnip juice agar and staling was consequently less, a smaller proportion of hyphae sought the plain substratum. Positively chemotropic reactions towards cane-sugar and glucose were also demonstrated but they were relatively weak. It may be inferred that the dominant influence governing the distribution of a parasitic fungus in its host, like that governing the distribution of saprophytic forms in culture, is negative chemotropism excited by the products of metabolism.

This inference is borne out by the observation of Robinson that the germ-tubes from the basidiospores of *Puccinia Malvacearum* failed to show any positively chemotropic curvature towards fragments of the host leaf.

**Hydrotropism.** Positive hydrotropism is probably effective in vegetative hyphae under appropriate conditions; thus Fulton found that, when spores of various species were grown on gelatine between two perforated mica plates with a relatively moist medium on one side and a relatively dry one on the other, they grew through the perforations towards the moist rather than towards the dry medium; hyphae of *Mucor Mucedo* grew through perforations in a mica plate from firm gelatine into water, but hyphae of *Rhizopus nigricans*, though they grew into the relatively moist gelatine near the holes, turned away from the fluid water below.

Germ-tubes of *Puccinia Malvacearum* were found by Robinson to grow from drops of water into the surrounding moist atmosphere, but on gelatine in moist air they tended to penetrate the gelatine. It would seem that the response varies, as is indeed to be expected, with the conditions of the fungus in question, but in some of these and perhaps other tropic reactions, there is at least a suspicion of the negatively chemotropic influence of staling substances.

Negative hydrotropism has been described for the sporangiophores of the Mucorales and, together with transpiration, has been held responsible for their divergence from one another when arising from a common point.

**Aerotropism and Osmotropism.** These factors have not so far been shown to play any important part in the directive growth of fungi.

**Phototropism.** A considerable number of fruit bodies are sensitive to light and by means of this reaction are able to adjust themselves in a position favourable to the distribution of their spores, the direction of light indicating the direction of open space.

Among the Mucorales the sporangiophores of various species of *Pilobolus*, *Mucor*, *Phycomyces* and no doubt of many other genera bend towards the light. In the genus *Pilobolus* the hemispherical sporangium is borne on an aseptate sporangiophore which develops a swelling or bulb just below the sporangium and another at its base. A new set of sporangia matures daily and is discharged in the morning or early afternoon. The young sporangiophores of the species studied by Jolivette[1] showed a recognizable phototropic curvature about 30 minutes after their exposure to light from a new direction; growth is apical and the tips, as they grow, bend towards the source of light. Curvature is arrested during the early stages of the formation of the sporangium and is resumed again when the subsporangial bulb is beginning to form; during the later stages of development curvature takes place just below the bulb. In this way the bulb and the terminal sporangium are pointed in the direction of the light and some accuracy of aim is secured. In a series of experiments, involving some 20,000 specimens, in which light reached the culture through apertures 1 cm. in diameter, nearly 90 °/ₒ of the sporangia hit either the aperture itself or the walls within 1 cm. on each side of it. When illuminated by two equal sources of light the sporangiophores point to either one or the other; this is obviously a useful adaptation as an intermediate aim would fail of its object. When the sources of light differed the sporangia were found to be shot off in larger numbers towards the light in which the proportion of blue rays was greater. The species[1] studied by Parr also responded more readily to blue or violet than to other rays, the presentation time gradually decreasing from red to violet although the sporangiophores were responsive to light from all regions of the spectrum.

Among Ascomycetes a positively phototropic response is found in the necks of the perithecia in *Sordaria* and many other Pyrenomycetes, and is sufficiently delicate to induce a zig-zag development of the neck if the direction of light is repeatedly changed. The asci of *Ascobolus immersus* and *A. furfuraceus* are also positively phototropic so that an appropriate direction is obtained for the ejection of their large spore mass.

As early as 1877 Brefeld recorded sensitiveness to light in the stipe of various *Coprini*, and found that normal pilei failed to develop in its

[1] Specific name not given.

absence. In *Coprinus niveus*, and *Coprinus curtus*, both coprophilous species, Buller found a well-marked positively phototropic response in young fruit bodies so that they push up from between or beneath the irregularities of the substratum. The stipe ceases to be phototropic when the pileus begins to expand and develops instead a negatively geotropic reaction; by these means the apex of the stipe is brought out into the light and the horizontal expansion of the pileus is ensured. A similar succession of reactions takes place in the development of the sporophore of *Lentinus lepideus*; the young stipe is positively phototropic and in the absence of light grows straight onwards without giving rise to a pileus; in a sufficiently strong illumination the pileus soon appears and, as its development proceeds, the positively phototropic response changes to a negatively geotropic one. *Amanita phalloides* and *A. crenulata* show, according to Streeter, a positively phototropic reaction even after the appearance of the pileus.

Corresponding reactions are to be expected in other pileate species growing on irregularly shaped substrata; on the other hand *Agaricus campestris*, growing on ground, is quite insensitive to light, negative geotropism being, under normal conditions, sufficient to secure the satisfactory adjustment of its parts.

A negatively phototropic reaction is very much less frequent than a positive response; Robinson found, however, that germ-tubes from the basidiospores of *Puccinia Malvacearum* and the conidia of *Botrytis sp.* turn away from a unilateral source of light, while those from the aecidiospores of *Puccinia Poarum* and from the conidia of *Peronospora parasitica* and *Penicillium glaucum* are indifferent.

**Phototaxis.** A phototactic reaction has been observed by Strasburger in the zoospores of *Chytridium vorax*, and by Wager in those of *Polyphagus Euglenae.* Both these species parasitize motile green organisms, *Chytridium* infecting *Chlamydococcus pluvialis*, and *Polyphagus*, *Euglena viridis.* The host organisms react to light since they obtain their carbon supplies by photosynthesis, and the phototactic reaction of the parasite brings it into the region where the hosts are to be found.

**Formative Influence of Light.** Irrespective of phototropic response the formative influence of light is important. Buller found that the sporophores of *Polyporus squamosus*, though quite irresponsive to the direction of light, fail to give rise to pilei unless illuminated. The same is true of many other Hymenomycetes, including those which, like *Lentinus lepideus* and the species of *Coprinus* mentioned above, are positively phototropic. In many cases only a brief period of illumination during the early stages of development is essential. Similarly, light appears to stimulate the development of ascocarps in certain Ascomycetes, and cultures may remain sterile in darkness or very dim light.

On the other hand Stevens and Hall found that the pycnidia of *Phyllosticta sp.* were irregularly scattered in continuous darkness but developed in regular concentric zones when exposed to the alternation of day and night; the rudiments of the pycnidia in this case are laid down mainly during the night.

**Geotropism.** The influence of gravity has been very inadequately studied among the lower fungi, little having been done since Sachs, in his *Lectures on the Physiology of Plants* in 1882, described the sporangiophores of *Mucor* and *Phycomyces* as bending up and the rooting hyphae down; he worked with filaments extending in all directions from a suspended piece of bread. Kny in 1881 reported that the mycelia of *Mucor Mucedo* and *Rhizopus nigricans* and also that of *Eurotium repens* are indifferent to gravity, and Miyoshi denied any geotropic reaction in his fungi.

Dawson in 1900 found that the stromata of *Poronia punctata*, a coprophilous pyrenomycete, show a well-marked negatively geotropic reaction from the earliest stages of their development and the same will probably prove true of other forms with upright stromata or with stalked fructifications like the Helvellales and many Pezizales.

Among the Hymenomycetes negative geotropism was first recorded by Sachs in 1860, and soon after by Hofmeister in 1863. In stipitate forms the stalk is always negatively geotropic as soon as the pileus develops, though, in some of the *Coprini* and in *Lentinus lepideus*, this may be preceded, as already seen, by a phase in which light is the directive influence.

Buller in 1909, and Streeter in the same year, showed that the geotropic response is of the nature of a gradual adjustment. The growing stalk swings beyond the vertical line, changes its direction and swings across it again, passing the vertical two or three times before it comes to rest. Curiously enough this method is followed in *Amanita crenulata*, in which the stalk reaches its full development in twenty-four hours, although there is not time for a complete adjustment, and the sporophore often comes to rest beyond the vertical line. The young stipe in *Amanita* elongates throughout its length until more than half grown; the zone of most rapid elongation is just below the pileus, it finally becomes narrower and narrower until growth ceases. Streeter was able to locate the perceptive region in the stipe, not in the pileus. *Amanita* showed a definite geotropic response when placed in a horizontal position for only one minute before being rotated on a klinostat. When stimulation lasted for a shorter period (15 or 30 seconds) no upward curvature ensued but a spiral curve was formed in the direction in which the klinostat moved. The latent period of young, vigorous specimens is about 40 minutes.

In the stemless Hymenomycetes, as in the stipitate forms, the orientation of the pileus takes place in response to gravity. This may be seen on any old stump where the bracket-shaped species of *Thelephora*, *Stereum*, *Polyporus* or *Polystictus* grow out as horizontal plates. Hasselbring found that the fruit bodies of *Polystictus cinnabarinus*, when rotated on a klinostat during their development, were no longer differentiated into a dorsal sterile and a ventral fertile region, but that hymenial tubes were formed over the whole surface.

Alike in the Hydnaceae, Agaricaceae and Polyporaceae the trama plates of the fertile region are positively geotropic. In stemless forms this reaction is responsible for the orientation of the hymenium; it may be particularly well seen in some of the Hydnaceae where the spines grow downward whatever the orientation of the sporophore, and it could doubtless be demonstrated also in *Tremellodon* among the Tremellaceae. In stipitate species it appears as a supplementary reaction coming into play if further adjustment is needed when the stipe is fully grown. The limitations of the method in the gill-bearing fungi are obvious for, if the pileus is oblique and the gills undergo much curvature, they become crowded together and interfere with spore dispersal.

## REACTIONS TO STIMULI: BIBLIOGRAPHY

1869 HOFMEISTER, W. Ueber die durch die Schwerkraft bestimmten Richtungen von Pflanzentheilen. Jahrb. wiss. Bot. iii, p. 92.

1877 BREFELD, O. Ueber die Bedeutung des Lichtes für die Entwickelung der Pilze. Sitzungsberichte der Gesellschaft Naturforschender Freunde zu Berlin, p. 127.

1878 STRASBURGER, E. Wirkung des Lichtes und der Wärme auf Schwärmsporen. Jena.

1879 VON SACHS, J. Ueber Ausschliessung der geotropischen und heliotropischen Krümmungen während des Wachsens. Arb. d. Bot. Inst. Würzburg, ii, p. 209. Or Lectures on the Physiology of Plants. Eng. Trans., 1887. Clarendon Press, Oxford, p. 700.

1881 DE BARY, A. Untersuchungen über die Peronosporeen und Saprolegnieen und die Grundlagen eines natürlichen Systems der Pilze. Beitr. zur Morph. und Phys. d. Pilze, iv, p. 85 of separate.

1881 KNY, L. Ueber den Einfluss äusserer Kräfte auf Anlegung von Sprossungen thallösser Gebilde. Sitz. Bot. Ver. Brandenburg, xxiii, p. 8.

1894 MIYOSHI, M. Über Chemotropismus der Pilze. Bot. Zeit. lii, p. 1.

1900 DAWSON, M. On the Biology of *Poronia punctata* (L.). Ann. Bot. xiv, p. 245.

1902 CLARK, J. F. On the toxic properties of some copper compounds, with special reference to Bordeaux Mixture. Bot. Gaz. xxxiii, p. 45.

1906 FULTON, H. R. Chemotropism of Fungi. Bot. Gaz. xli, p. 81.

1906 PFEFFER, W. The Physiology of Plants. Eng. Trans. Clar. Press, Oxford, iii, p. 183.

1907 HASSELBRING, H. Gravity as a Form-Stimulus in Fungi. Bot. Gaz. xliii, p. 251.

1908 BALLS, W. L. Temperature and Growth. Ann. Bot. xxii, p. 558.

1909 BULLER, A. H. R. Researches on Fungi. Longmans, Green & Co., London.

1909 STEVENS, F. L. and HALL, J. G. Variations of Fungi due to Environment. Bot. Gaz. xlviii, p. 1.

1909 STREETER, S. G. The Influence of Gravity on the Direction of Growth of *Amanita*. Bot. Gaz. xlviii, p. 414.

1914 JOLIVETTE, H. D. M. Studies in the Reactions of *Pilobolus* to Light Stimuli. Bot. Gaz. lvii, p. 89.

1914 ROBINSON, W. Some Experiments on the Effect of External Stimuli on the Sporidia of *Puccinia malvacearum* (Mont.). Ann. Bot. xxviii, p. 331.

1914 WAGER, H. Movements of Aquatic Micro-Organisms in Response to External Forces. The Naturalist, No. 689, p. 171.

1916 GRAVES, A. H. Chemotropism in *Rhizopus nigricans*. Bot. Gaz. lxii, p. 337.

1918 PARR, R. The Response of *Pilobolus* to Light. Ann. Bot. xxxii, p. 177.

# CHAPTER II

## ASCOMYCETES

**The Ascomycetes** include over 15,000 species, all of which, excepting only the yeasts, possess a well-developed mycelium of richly branched and septate hyphae. The cells of the mycelium may be uninucleate, as in the Erysiphaceae and species of *Chaetomium* and *Sordaria*, or they may contain a few or several nuclei; energetic growth and a rapid succession of nuclear divisions often cause the nuclei of multinucleate elements to be arranged in pairs.

Multiplication may take place by means of conidia, oidia, or chlamydospores, but the characteristic method of reproduction is by **ascospores** or spores produced in the interior of a mother-cell or **ascus**.

**The Ascospores.** In the large majority of species the ascospores are elliptical in outline, but they may be spherical or globose, as in *Ascodesmis nigricans*, long and narrow (filiform) as in species of *Cordyceps* or *Claviceps*, or of intermediate form. They contain a densely granular or reticulate cytoplasm in which one or more oil drops are usually present. The epispore may be smooth or variously sculptured; in several cases it is verrucose, in others reticulate; the latter arrangement is particularly well seen in *Ascobolus furfuraceus*.

All ascospores are colourless when first formed; they may remain so when ripe, or may assume a variety of colours.

The young spore is unicellular and uninucleate[1]. Before it is set free the nucleus may divide, and this division is frequently accompanied by wall-formation so that, like the asexual spore of *Pellia*, the ascospore may be regarded as undergoing premature germination. Divisions may take place in three dimensions, so that the spore is **muriform** (fig. 1), or the septa may be all in one plane so that it consists of a row of cells (fig. 2 *a*).

The multicellular spore thus produced may break up into its constituent cells, which proceed with their development independently, or it may function as a single structure.

In examining any of these characters or in measuring the size of the spore it is essential to deal with fully mature examples since the distinctive sculpturing, pigmentation, etc., may not appear till late in development. Spores which have escaped naturally from the ascus may be used with safety.

[1] Lewton Brain (*Ann. Bot.* 1901) describes the spores of *Cordyceps ophioglossoides* as multinucleate from their first inception.

**The Ascus.** The ascus or mother-cell of the spores is a spherical, oval, club-shaped, or almost cylindrical organ with a narrow, more or less elongated

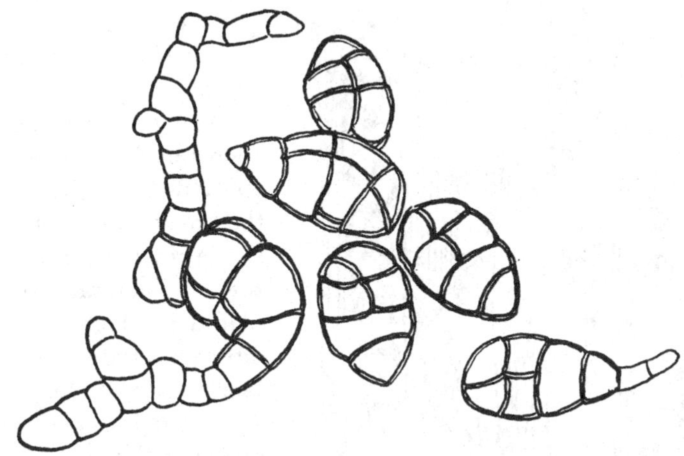

Fig. 1. *Pleospora sp.*; germinating spores, × 1000.

base. When moderately young it contains a single nucleus which undergoes three karyokinetic divisions giving rise to eight daughter nuclei (fig. 3). Asci of the short, stout type are full of dense cytoplasm; in the relatively cylindrical forms the ends are usually vacuolate, but a broad, granular belt fills the middle region and contains the nuclei. The spores are cut out by free cell-formation so that a certain amount of cytoplasm remains outside them, constituting the **epiplasm**. It becomes charged with glycogen and other food substances, and is a source of supply to the developing spores.

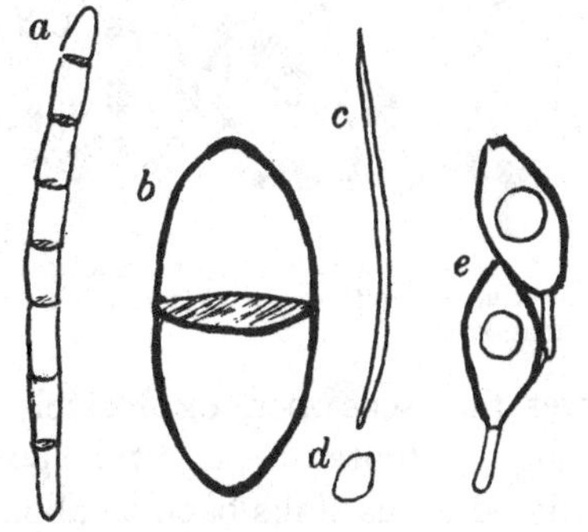


Fig. 2. Spores of *a. Geoglossum difforme* Fr.; *b. Delitschia furfuracea* Niessl; *c. Rhytisma acerinum* (Pers.) Fr.; *d. Chaetomium Kuntzeanum* Zopf; *e. Podospora minuta* (Fuck.) Wint.; × 500.

In the vast majority of cases the mature ascus contains eight spores, but in a certain number of species, though eight nuclei are produced, only one or a few are utilized as centres of spore-formation. Thus a single spore is sometimes developed in *Tuber*, two is the regular number in *Phyllactinia*, and four in many of the Laboulbeniales; in *Bulgaria polymorpha* eight spores are initiated but only four reach maturity. In *Endomyces* it is probable that the ascus nucleus divides only twice, and the four spores utilize all the available nuclei. On the other hand additional nuclear divisions sometimes precede spore-formation so that the number of spores is increased. There are sixteen or thirty-two in the species of *Rhyparobius*, sixteen to sixty-four in *Podospora pleiospora*, and in *Podospora curvicolla* one hundred and twenty-eight.

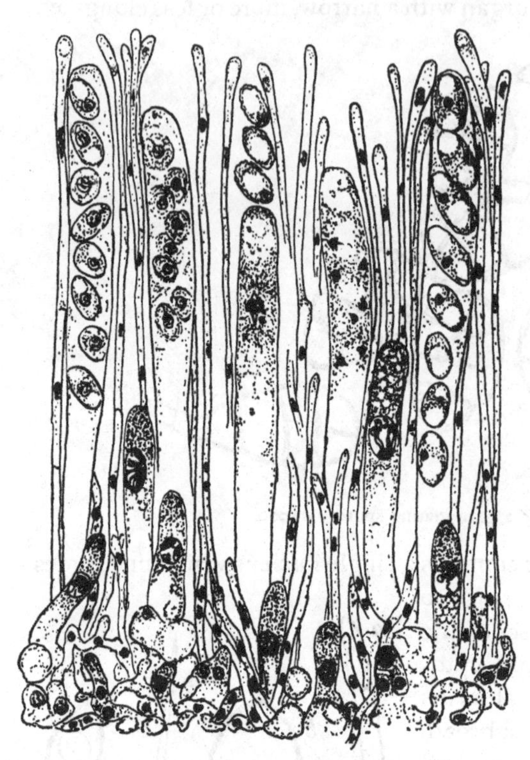

Fig. 3. *Humaria rutilans* (Fr.) Sacc.; hymenial layer showing asci and paraphyses in various stages of development, × 400.

The arrangement of the spores in the ascus is usually constant for a given species; they may be **uniseriate**, in a single row; **biseriate**, in two rows somewhat irregularly placed, in which case they are often uniseriate when young; **fasciculate** if long narrow spores are arranged in a parallel bundle; or **inordinate** if they show no regular arrangement. In the simplest forms the spores are liberated by the decay of the ascus walls, and, where a definite fruit body is present, they remain for a time enclosed by its outer layers. In other cases, including the majority of Discomycetes and Pyrenomycetes, the ascus opens explosively either by an irregular tear or by dehiscence along a definite line, and the spores are shot out in a jet of liquid while the deflated ascus sinks back to about half its size (figs. 4, 5). In forms with an explosive mechanism the ascus often elongates considerably during the latter part of its development; the spores are arranged at the upper end and either float suspended in the fluid contents of the ascus, or are attached to the apex and to one another by cytoplasmic strands (*Sordaria*, *Podospora*, fig. 2 *e*).

The explosive ejection of spores from different asci may be simultaneous or successive; in a certain number of forms with an exposed hymenium of parallel asci (Pezizales, Helvellales, Exoascales) successive discharge takes place under moist conditions, but any disturbance leading to rapid loss of water causes the simultaneous liberation of a number of spore masses, so that the cloud of spores is visible even to the naked eye. This phenomenon, which is known as "puffing," may be brought about under conditions of moderate dryness, such as occur out of doors on a fine autumn day, by shaking the fructifications, or even by currents of air set up by walking past them. It can be initiated, as de Bary pointed out, when ripe

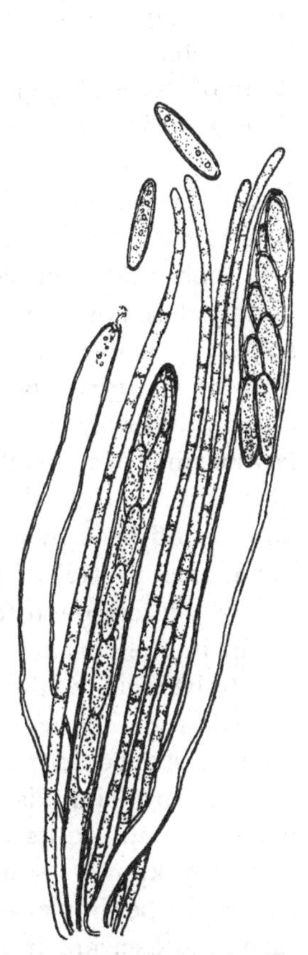

Fig. 4. *Mitrula laricina* Mass.; development and ejection of biseriate spores, × 600.

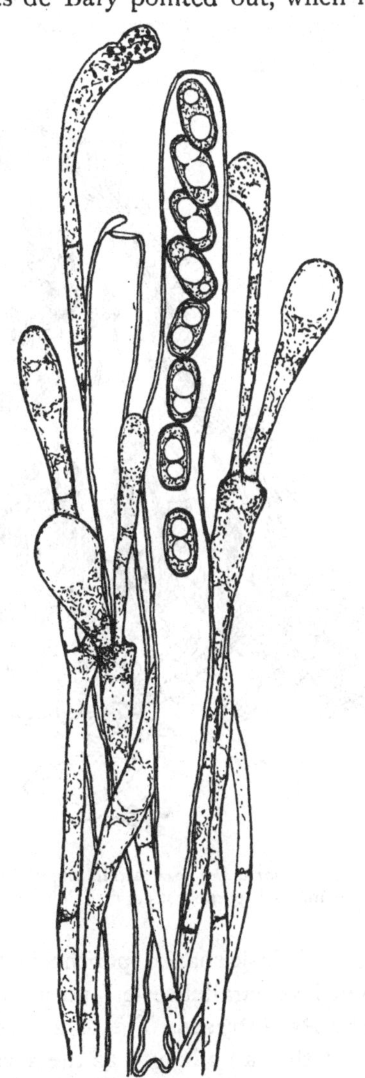

Fig. 5. *Sepultaria coronaria* Mass.; uniseriate spores; ascus opening by a lid; branched, septate, clavate paraphyses; × 600.

isolated asci lying in water are suddenly exposed to the action of glycerine or alcohol, and is clearly due to alterations of tension affecting a number of asci at about the same stage of development. After the fructification has puffed once or a few times a rest of some hours during which fresh asci reach maturity is necessary before the phenomenon can be repeated.

**The Ascocarp.** In a few genera, asci are developed singly on the mycelium, but in the great majority of cases they arise in closely associated groups, and are surrounded by a protective wall or **peridium** of sterile filaments so that a definite fructification, the **ascocarp** or **ascophore**, is formed. This structure is usually more or less spherical and completely closed in the young stages; it may retain this form at maturity, opening only by the decay or irregular splitting of its walls, it may assume a flask-shaped outline and open by a terminal pore or **ostiole** (fig. 6), or it may spread out to form a cup in the concavity of which the asci are fully exposed (fig. 7). To the closed or flask-shaped forms the term **perithecium** is applied, the cup and its variants are known as **apothecia.** In the simpler ascocarps the asci are irregularly scattered; in the apothecia and flask-shaped perithecia they are regularly arranged, forming a more or less parallel series and intermingled with paraphyses.

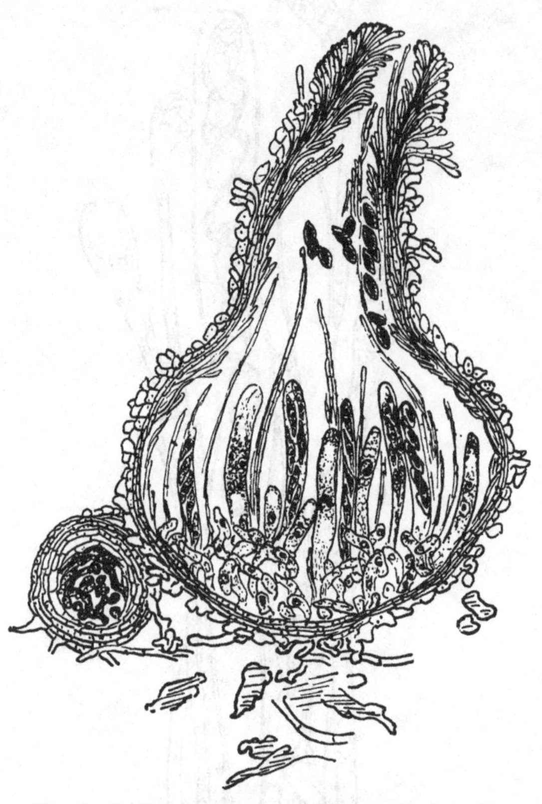

Fig. 6. *Sordaria sp.*; ascocarp in longitudinal section showing asci, paraphyses and periphyses; ×400.

**The Paraphyses.** The paraphyses of the Ascomycetes are slender hairs, of about the same length as the asci; they usually develop earlier than the latter, and have a protective and possibly a nutritive function; in the simpler ascocarps their place is taken by the inner layers of the sheath. The paraphyses are often **clavate** or club-shaped in form with a rather swollen tip (fig. 5), sometimes cylindrical (fig. 4) or pointed (**lanceolate**), or sometimes with curled or twisted ends; they may be simple or branched, septate or continuous, hyaline or provided with coloured contents; orange red granules are common and often give a brilliant tinge to the whole hymenium. In *Desmotascus*[1], a pyrenomycetous fungus parasitic on *Bromelia*, the paraphyses are replaced by a thin-walled pseudoparenchyma recalling the arrangement in the higher Plectomycetes.

**The Peridium.** The peridium or wall of the ascocarp is a weft of sterile hyphae in which the individual filaments are sometimes clearly distinguished, sometimes closely interwoven to form a pseudoparenchyma; the walls of the outer cells are sometimes considerably thickened and may be variously pigmented; in many cases they give rise to hairs (*Lachnea*, *Chaetomium*) or other appendages (Erysiphaceae).

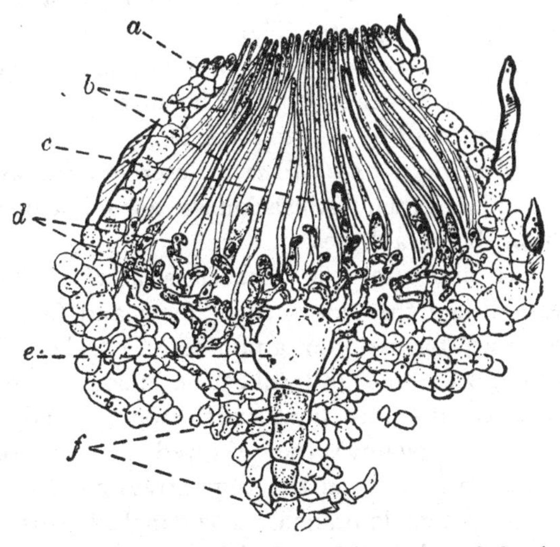


Fig. 7. *Lachnea stercorea* (Pers.) Gill.; ascocarp in longitudinal section showing young asci and paraphyses; the oogonium is still recognizable; × 160. *a.* sheath, *b.* paraphyses, *c.* ascus, *d.* ascogenous hyphae, *e.* oogonium, *f.* stalk of archicarp.

**Alternation of Generations.** The Ascomycetes show a well-marked alternation of generations modified by the wide occurrence of apogamy in the group.

In forms where fertilization is still retained the mycelium arising from the germination of the ascospores gives rise to the sexual organs. These show a considerable variety of form and in the vast majority of cases are clearly differentiated as male and female structures. The male branch consists of an antheridial hypha or **stalk** and a terminal **antheridium** which, in some of the more complex forms, is detached and carried by the wind or otherwise as a **spermatium**. The female branch or **archicarp** possesses a **stalk** of one or more cells, bearing the **oogonium**; sometimes a terminal **trichogyne** is present by means of which connection is established between the oogonium and antheridium. The term **ascogonium** has been used at various times as the equivalent of archicarp or oogonium, or to indicate the oogonium after fertilization, but it does not appear essential in any of these senses. In certain cases where the walls are partly broken down between several cells in the middle of the archicarp, the term **oogonial region** is applied to them. An archicarp of this type is sometimes known as a **scolecite.** The gametangia each contain one or several nuclei, but the contents are in no case rounded off as independent gametes. The

[1] 1919, Stevens, F. L., "Perithecia with an Interfascicular Pseudoparenchyma," *Bot. Gazette*, lxvii, p. 422.

contents of the antheridium enter the oogonium which, as a result of this association, gives rise in nearly every case, with or without preliminary septation, to a number of filaments, the **ascogenous hyphae**, from the tips of which asci grow out. The ascogenous hyphae thus constitute the sporophyte while the vegetative mycelium, on which the sexual organs are borne, is the gametophyte. The gametophyte gives rise to the peridium and the paraphyses, and on it the various accessory spores are produced.

**Early Investigators.** The history of the minute study of the Ascomycetes may be said to have begun in 1791 when Builliard, in his *Histoire des Champignons de France*, described the asci as female organs, and suggested that their fertilization was accomplished by some substance emanating from the paraphyses.

In and after 1863 the classical researches of de Bary and his pupils established the existence of male and female organs at the beginning of ascocarp formation in a number of species. They brought forward evidence of the occurrence of fertilization in some cases and in others of development without an antheridium (parthenogenesis) or without either male or female organs (apogamy), they showed that the paraphyses and sheath of the ascocarp arise from the vegetative mycelium and the ascogenous hyphae from the female branch. The conclusions reached by de Bary are summed up in the fifth chapter of his book on the *Comparative Morphology and Biology of the Fungi, Mycetozoa and Bacteria*, first published in 1884, and they have been largely confirmed by subsequent investigation.

In 1881 Eidam observed the formation of the ascus in the very simple genus *Eremascus*, where it arises from two separate filaments which become intertwined and fuse at their tips.

De Bary's views were extensively criticized, especially by van Tieghem and Brefeld, who both denied the occurrence of sexuality in the group. These and other writers sought to explain the antheridial filament as part of the sheath, and the archicarp as a precocious ascogenous hypha, or, in certain lichens, as a boring or a respiratory organ. Brefeld was the author of a scheme of classification, which, if too rigid to endure the test of subsequent work, was at least exceedingly convenient. With it, and especially in the text book of his disciple von Tavel (1892), his view that the higher fungi lacked sexuality was widely disseminated.

**Cytology.** In the meantime considerable advances were being made in the study of cytology and of the cytological methods necessary for the examination of minute forms. De Bary, in 1863, had recognized the presence of a single or definitive nucleus in the immature ascus of *Pyronema confluens* and some other species, and the successive appearance as development proceeded of two, four, and eight nuclei. He found that the eight nuclei lay at more or less equal distances apart, and that each became surrounded by a mass of cytoplasm forming the primordium of a spore. In 1879 Schmitz observed nuclei in the vegetative cells of several Ascomycetes, and in 1893 Gjurasin in *Peziza vesiculosa* recognized that the divisions in the ascus are karyokinetic.

**The Fusion in the Ascus.** In 1894, Dangeard showed in *Peziza vesiculosa* and other forms with a well-developed fructification, that the ascus at its first inception is binucleate and that the two nuclei subsequently unite to form the definitive nucleus of de Bary. He at first believed that the ascus was produced in these cases, as in *Eremascus*, by the fusion of two independent filaments, but he was soon able to ascertain that the young ascus arises, not by the union of two separate hyphae, but from the penultimate cell of a single recurved filament. The apex of this filament receives two nuclei which undergo a simultaneous karyokinesis, so that four are formed. One of these lies at the tip of the hypha, and one passes back into the lower part, while the other two lie in the curved portion, and become separated from their sister nuclei by transverse walls (fig. 8). The terminal cell of the hypha thus contains a single nucleus and the penultimate cell is regularly binucleate; it grows out laterally to form the ascus, and its two nuclei fuse soon after they come together.

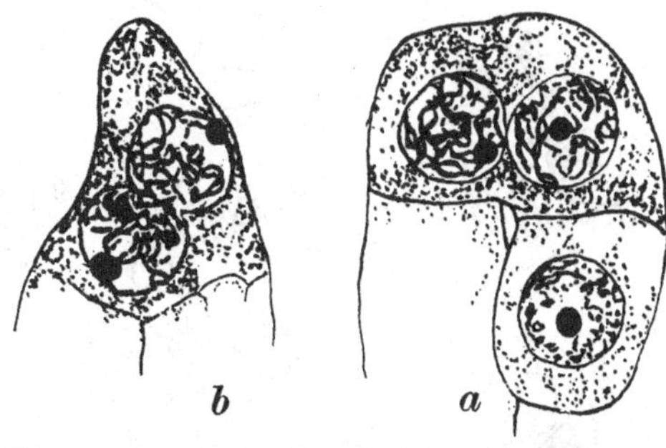


Fig. 8. *Humaria rutilans* (Fr.) Sacc.; *a*. ascogenous cell containing two nuclei cut off from the uninucleate terminal cell and stalk; *b*. fusion in the ascus, the nuclei are just passing into synapsis; both × 1875.

The fusion in the ascus was the first nuclear fusion observed among Ascomycetes and its discoverer, Dangeard, accepting it as a sexual process, regarded the ascus as an egg, and the sexual apparatus described by de Bary and his pupils as at most vestigial. It is probable that this view was influenced by his first opinion that the ascus arose from the fusion of two *separate* filaments, but it was not modified by his subsequent discovery of the true process.

**Fertilization.** In the following year (1895) both de Bary's observations and those of Dangeard were confirmed by Harper, working on the common mildew *Sphaerotheca Humuli*[1]. Harper saw the development of a uninucleate oogonium and a uninucleate antheridium. He observed the passage of the male nucleus into the oogonium and the fusion there of the sexual nuclei. After fertilization the oogonium underwent septation, giving rise to a row of cells of which the penultimate contained two nuclei, and, after these had fused, developed into the ascus (fig. 9). He thus demonstrated that in this fungus there is a normal fusion of male and female nuclei, followed by a

[1] *Sphaerotheca Humuli* (DC.) Burr. = *Sphaerotheca Castagnei* Lév.

second fusion in the ascus, and in the following year he recorded the same process in *Erysiphe Polygoni*.

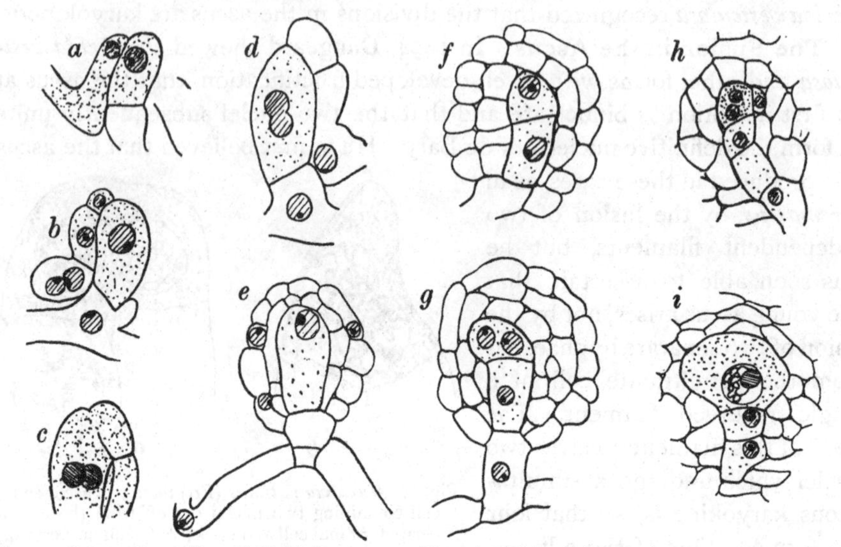


Fig. 9. *Sphaerotheca Humuli* (DC.) Burr.; *a.* and *b.* antheridium and oogonium; *c.* entrance of male nucleus; *d.* fusion in oogonium, antheridium without nucleus; *e.* fusion nucleus in oogonium; *f.* and *g.* septation of oogonium; *h.* two nuclei in ascus; *i.* ascus after nuclear fusion; after Harper.

In 1900, Harper observed fertilization in the oogonium of *Pyronema confluens*; here, however, the gametangia are both multinucleate and fertilization consists of the fusion in pairs of a large number of male and female nuclei. Many asci are produced, each from a recurved filament in the penultimate cell of which a second nuclear fusion occurs.

**Development of the Ascus.** Though the young ascus, so far as observation goes, is almost invariably binucleate and the seat of a nuclear fusion, the details of its formation are not always the same. In 1905 Maire showed that in *Galactinia succosa* and occasionally in other forms a series of three or four binucleate cells is produced at the end of each ascogenous hypha. The nuclei of the terminal cell undergo simultaneous division and two are cut off at the apex by a transverse wall. These fuse and the cell containing them becomes the ascus. Harper in *Erysiphe*, and other authors in various other plectomycetous fungi have found that any cell of an ascogenous hypha, if it contains two nuclei, may give rise to an ascus without preliminary nuclear division.

In some of those species in which the ascus is derived from a binucleate penultimate cell the curvature of the hypha is so great that the uninucleate terminal cell lies in contact with the stalk cell of the ascus. When this happens the terminal and stalk cells sometimes fuse, a nucleus wanders from one to the other, and the cell thus provided with two nuclei grows out as a continuation of the ascogenous hypha, and gives rise to fresh asci (fig. 10). This process was first recorded in 1908 for *Humaria rutilans* and has since been observed by McCubbin in *Helvella elastica*, by Carruthers in *Helvella crispa*, and by Claussen in *Pyronema confluens*. It suggests either that some advantage is to be derived from an absence of relationship between the nuclei which fuse in the ascus, or that a scheme of rigid nuclear economy is in force. The former hypothesis is somewhat weakened by the fact that no means of avoiding close relationship appear to exist in the Plectomycetes.

**Meiosis.** Very soon after the ascus is cut off, preparations are made for the first nuclear division, which was shown in 1905 by Guilliermond, Harper and Maire, working independently on various fungi, to be heterotype, and to be followed by a second which is homotype in character; their

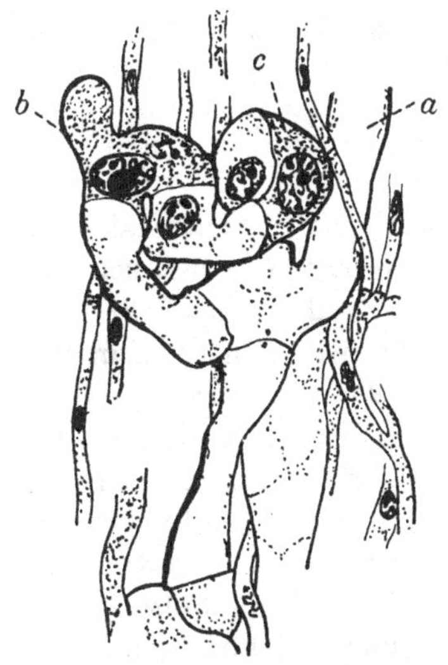


Fig. 10. *Humaria rutilans* (Fr.) Sacc.; an ascus (*a*) the terminal cell connected with which has continued its growth and given rise to another ascus (*b*) from the terminal cell of which a third ascus (*c*) has arisen, × 1250.

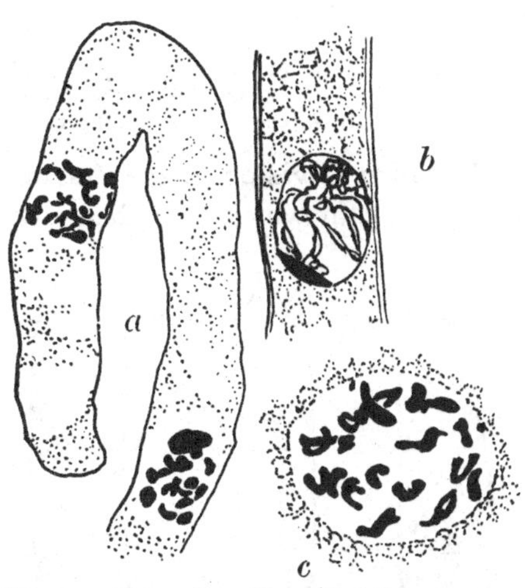


Fig. 11. *Humaria rutilans* (Fr.) Sacc.; *a.* ascogenous hypha showing sixteen chromosomes in each nucleus, × 1950; *b.* fusion nucleus of ascus passing out of synapsis, × 1300; *c.* fusion nucleus of ascus showing sixteen gemini, × 1950.

observations have since been widely confirmed by a number of investigators, and synapsis, the second contraction and the formation of typical gemini have been seen, as well as the reduction of the chromosome number.

Thus in *Humaria rutilans*, which the exceptionally large nuclei render a convenient subject of study, each of the fusing nuclei possesses sixteen chromosomes (fig. 11 *a*), so that the definitive nucleus has thirty-two; after meiosis is complete sixteen can be counted in each daughter nucleus. This fungus is somewhat exceptional in that synapsis begins separately in each of the two nuclei of the young ascus before they fuse (fig. 8 *b*) indicating that each of them is already diploid, the product of a sexual fusion, and capable of independent meiosis. In *Helvella crispa* synapsis takes place after the nuclei have fused, but the two spiremes contract in separate masses at opposite ends of the nuclear area (fig. 12)[1]. In other investigated species only one synaptic knot is formed.

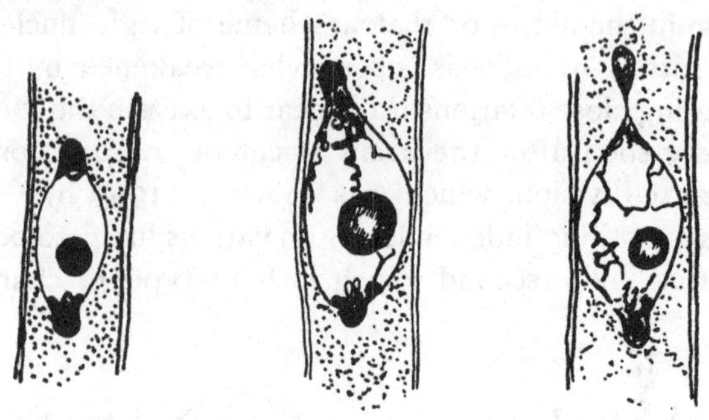

Fig. 12. *Helvella crispa* Fr.; fusion nuclei in ascus showing the contraction of the chromatin in two separate masses, × 2000; after Carruthers.

**The third Division in the Ascus.** The meiotic phase is followed by a third division in which a further change in the chromosome number has been described. This was first recorded by Maire, in 1905, in the case of *Morchella esculenta*, and, with certain variations in detail, in two other Discomycetes. In *M. esculenta* there are eight chromosomes in the first telophase, and four in the third, and also in the subsequent nuclear division which takes place in the spore. The next case to be brought forward was *Humaria rutilans*, in 1907, and two more (*Ascobolus furfuraceus* and *Pyronema confluens*) were recorded by Dangeard in the same year (fig. 13). It was soon after suggested (Fraser, 1908) that the halving of the chromosome number which had been observed in these species constituted a second reduction phase, bearing the same relation to the fusion in the ascus that meiosis bears to the fusion of the sexual nuclei. For this reduction the term **brachymeiosis**

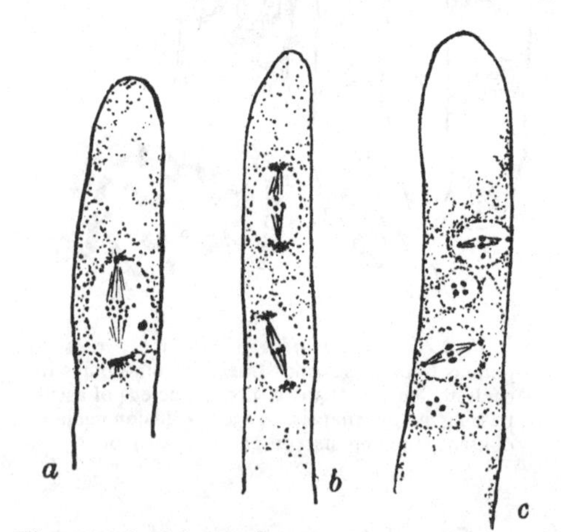


Fig. 13. *Ascobolus furfuraceus* Pers.; *a*. early anaphase of the first division in ascus showing 14 of 16 daughter chromosomes; *b*. metaphase of the second division showing four chromosomes; *c*. third division showing four; after Dangeard.

[1] This arrangement may occasionally be found in *Humaria rutilans*, and occasionally also the two nuclei of the young ascus proceed as far as the second contraction without undergoing fusion.

was proposed. The occurrence of a brachymeiotic reduction has since been observed in several other fungi, and has also been in several cases denied.

**Chromosome Association.** There are a number of fungi, of which *Phyllactinia Corylea* is perhaps the most fully studied, in which no change in the chromosome number takes place throughout the life-history. In *Phyllactinia*, Harper showed in 1905 that the chromosomes remain visible in strands attached to the central body throughout the resting stages. In each of the nuclei of the developing ascus eight such strands can be clearly seen (fig. 14), and their association in pairs can be followed, so that they appear as eight strands in the spireme stage, and as eight chromosomes on the spindle. In the smaller sexual nuclei Harper found a similar arrangement. It would thus appear that in this fungus chromosome association follows directly on nuclear fusion, so that the fusion of two nuclei with *n* univalent strands produces not a nucleus with 2*n* strands, but one with *n* bivalent strands. If this is so in the oogonium the nuclei which fuse in the ascus will each possess *n* bivalent chromosomes, and the definitive nucleus will show not 4*n* univalent chromosomes but *n* which are quadrivalent. In the same way neither meiosis nor brachymeiosis will affect the chromosome number, but will affect only the valency of the chromosomes. This is the case in *Phyllactinia*, where in all stages of the three divisions in the ascus, eight chromosomes are found. It must be noted, however, that in cases where no fusion occurred in the life-history except that which immediately precedes meiosis, the chromosome number would similarly remain unchanged; the latter interpretation has been urged as evidence that forms in which the chromosome number is unaltered throughout the life-history are therefore necessarily without a fusion in the oogonium.

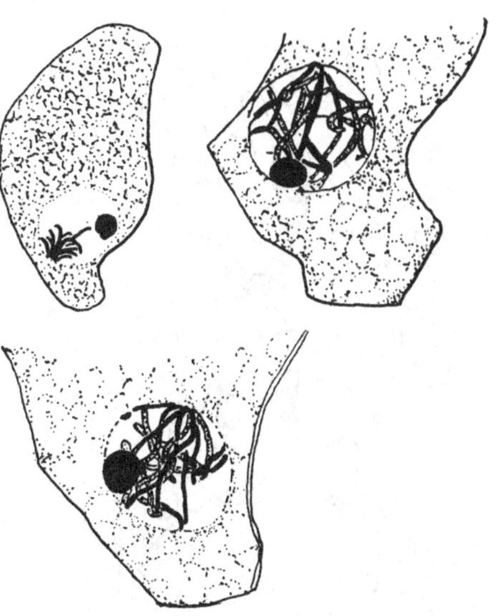

Fig. 14. *Phyllactinia Corylea* (Pers.) Karst.; fusion nuclei in ascus showing eight chromatin strands attached at a common point; after Harper.

**The Theory of a Single Nuclear Fusion.** The possibility that the fusion in the ascus is the only nuclear fusion in the life-history of the Ascomycetes, and represents the postponed union of sexual nuclei which had become associated in fertilization was first mooted by Raciborski in 1895, in a letter to Professor Harper, and was published in 1896. This view received a considerable impetus in 1907 and 1912, from the work of Claussen on *Pyronema confluens.* According to the latter author the male nuclei enter the oogonium and pair with the female nuclei, but do not fuse with them. The sexual nuclei pass in pairs into the ascogenous hyphae (fig. 15), and eventually they or their descendants fuse in the ascus. For Claussen, therefore, there is a single fusion (that in the ascus) and a single reduction (the meiotic) in the life-history of *Pyronema.* His interpretation has been followed by Schikorra on *Monascus,* by Faull on *Laboulbenia,* and by Ramlow on *Ascobolus immersus* and *Ascophanus carneus.* The last-named author figures certain "fusions" in the oogonial region of his fungi, but regards them as pathological.

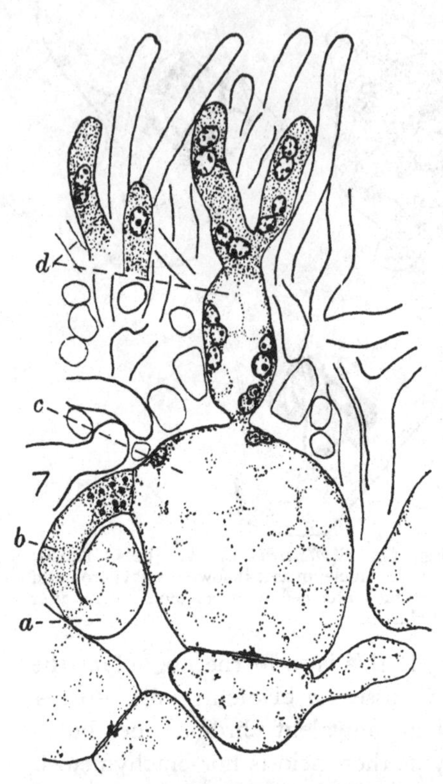


Fig. 15. *Pyronema confluens*; sexual apparatus and paired nuclei in the ascogenous hyphae; *a.* antheridium; *b.* trichogyne; *c.* oogonium; *d.* ascogenous hyphae; × 1040; after Claussen.

Claussen's hypothesis demands that the attraction between the sexual nuclei, though not sufficient to cause fusion, yet holds them together in the multinucleate ascogenous hyphae and is transmitted to their descendants when nuclear division has occurred[1]. It is based on three chief grounds:

(1) the failure to observe fusions in the oogonium or their interpretation, if found, as pathological phenomena;

(2) the recognition of as many chromosomes in the third division in the ascus as in the first;

(3) the observation of paired nuclei in the ascogenous hyphae.

The first of these grounds has a mainly negative value; in regard to the second, further investigation is very much to be desired; the statement of the chromosome number without figures is of little value, nor is any figure of the third division significant except that of the late anaphase; in the earlier stages chromosomes are scattered about the spindle, so that there is no criterion by which a metaphase showing eight undivided chromosomes can be distinguished with surety from an early anaphase showing two sets of

[1] A comparison is sometimes instituted between the sporophyte of the Ascomycetes and that of the rusts. In the rusts, however, each pair of nuclei is enclosed in a separate cell and the hypothesis of nuclear attraction throughout the vegetative phase is accordingly not required.

four. Further the adhesion of chromosomes already described for *Phyllactinia* must not be forgotten.

The occurrence of paired nuclei in the ascogenous hyphae was thus the most important evidence in favour of Claussen's view until in 1916 Welsford showed that the nuclei even of gametophytic, multinucleate hyphae are habitually paired if rapid growth and division are taking place; this is due to the fact that mitoses follow one another so rapidly that the daughter nuclei of any particular division have not time to move apart, before they themselves divide. Such paired nuclei have often previously been figured though without attracting special attention; excellent examples are to be seen in Nichols' paper on the Pyrenomycetes[1] or in Ramlow's more recent work on *Ascophanus carneus*[2]. It seems hardly possible to place a different interpretation on the nuclei which lie close together in the ascogenous hyphae.

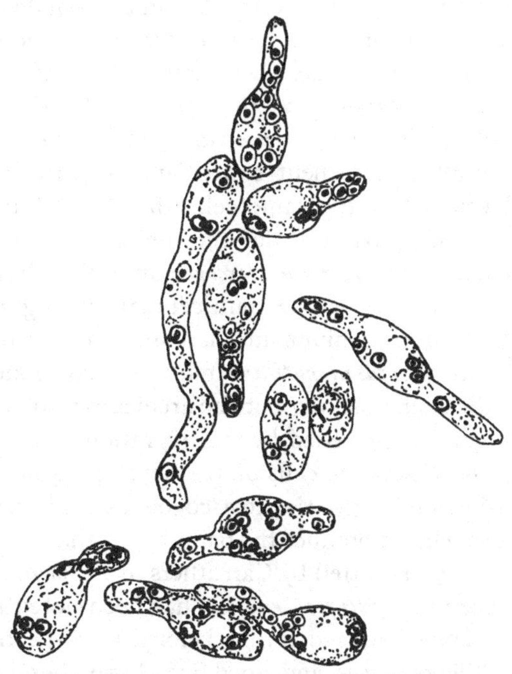

Fig. 16. *Ascophanus carneus* Pers.; germinating spores with paired nuclei in the germ-tubes, ×450; after Ramlow.

According to our present knowledge of the cytology of the Ascomycetes there are two nuclear fusions in the life-history of these plants.

**The Significance of the Fusion in the Ascus.** If this be the case it remains to consider the significance of the fusion in the ascus. The presence of more than one nucleus in this cell, destined to be one of the largest in the life-cycle of the fungus, is hardly surprising especially in coenocytic forms. In uninucleate species it forms part, as Harper pointed out in 1905, of the quantitative adjustment frequently observed between cytoplasm and nuclear material. But this nucleo-cytoplasmic relation does not explain why fusion should take place between the nuclei concerned or why they should be regularly two in number[3]. It is possible that, crowded as they are in the newly constituted ascus, the nuclei merely flow together as they make ready for the prophases of division. Whatever may have been the determining

[1] *Bot. Gaz.* 1896, xxii, pl. xv, fig. 32. [2] *Myc. Centralbl.* 1915, pl. i, fig. 12.

[3] In *Humaria rutilans*, however, and doubtless in other forms, young trinucleate, and quadrinucleate asci are found.

cause of the second fusion in the ancestors of our present Ascomycetes, it is clear that in the forms now extant the existence of a second reduction must be an important factor; in this respect the organism may well be moving in a vicious circle.

**Pseudapogamy.** The above, at any rate, seems to be the case in respect of the much more important fusion of the sexual nuclei. In *Humaria granulata* no antheridium is developed, and the female nuclei, as recorded by Blackman and Fraser in 1906, fuse in pairs in the oogonium before passing into the ascogenous hyphae. The same state of affairs was observed by Welsford in *Ascobolus furfuraceus* in 1907, and by Cutting in *Ascophanus carneus* and Dale in *Eurotium repens* in 1909; it is difficult to see how any important physiological benefit can ensue from the union of closely related nuclei developed in the same cell; the determining cause would seem to be the need of preparation for the meiotic phase already established in the life-history. In *Lachnea stercorea* an antheridium is present and fuses with the terminal cell of the multicellular trichogyne, but the male nuclei never reach the oogonium, and here also the female nuclei unite in pairs.

In *Humaria rutilans* matters have gone still further; not only is the antheridium lacking but the archicarp also is not developed. Nuclear fusion, sometimes preceded by the migration of one of the fusing nuclei, takes place in the vegetative cells of the developing apothecium. The cells containing fusion nuclei give rise to ascogenous hyphae, while those in which fusion has not occurred produce the paraphyses and the sheath. A similar state of affairs has been reported by Carruthers in *Helvella crispa*, and evidence of its occurrence in *Polystigma rubrum* has been noted by Blackman and Welsford.

These reduced forms, belonging respectively to de Bary's categories of parthenogenesis and apogamy, have thus proved to be **pseudapogamous** in the sense of Farmer and Digby[1], since in them normal fertilization is replaced by the union in pairs of female or vegetative nuclei. Meiosis takes place as usual.

**Spore-Formation.** After the third division in the ascus, preparations for spore-formation begin. This stage was first described in detail by Harper in 1895, and was subsequently elucidated by him in other papers, and especially in his very full study of the mildews in 1905. As the third mitosis comes to an end the eight daughter nuclei, or those of them about which spore-formation is to take place, become pear-shaped, a beak being pushed or pulled out from each; the centrosome lies at the tip of the beak, and from it spread the astral rays, to the activity of which Harper is inclined to attribute the formation of the beak. As development proceeds, these rays become folded over so that they extend past the nucleus, and Harper describes them as combining side by side to form a continuous, broad, umbrella-like mem-

[1] *Ann. Bot.* 1907, xxi, p. 191.

brane which gradually closes in to produce, by further marginal growth, the ellipsoidal plasma membrane of the spore. In this way the whole body of the spore is cut out from the undifferentiated cytoplasm of the ascus by a process of free cell formation, and its membrane is formed by the fusion of the astral rays.

A different account of the development of the membrane was given by Faull in 1905. According to his investigations the spore is cut out by the gradual differentiation from the centrosome downwards of a limiting layer of hyaline or finely granular cytoplasm, in the production of which the astral radiations play no part.

In many of the Discomycetes, however, the importance of the aster in spore-formation is very marked, and the spore is outlined by radiations passing out from the centrosome (fig. 17). These radiations, in *Humaria*

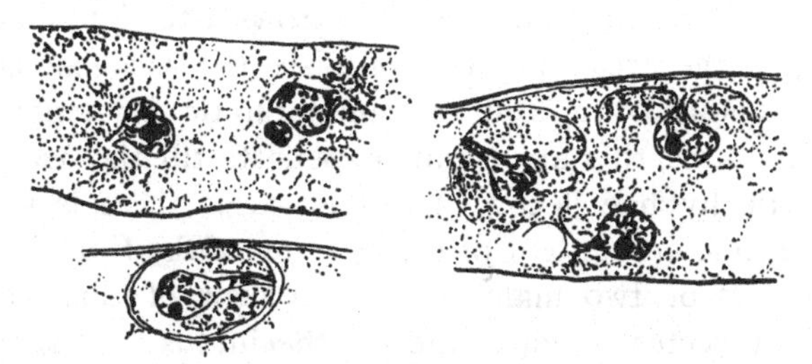

Fig. 17. *Humaria rutilans* (Fr.) Sacc.; stages of spore formation, × 1875.

*rutilans*, *Peziza vesiculosa*, or *Lachnea stercorea*, doubtless indicate the paths of altered substances emanating from the centrosome as a centre of activity, and flowing back past the nucleus as the developing beak pushes into the cytoplasm. As these substances increase a membrane is formed, and the spore, or the part of it near the centrosome, is cut out. In the delimitation of the region remote from the centrosome the vacuoles in the cytoplasm of the ascus may take part. These vacuoles are especially plentiful in *Ascobolus furfuraceus*, and in this fungus their share in spore-formation is important. After the spore is delimited the differentiation of its wall is largely due to the epiplasm, and is not complete till the ascus is almost ripe.

**Phylogeny.** The recognition of the specialized character of the ascus has led to a general assumption of the monophyletic origin of the Ascomycetes, and speculation as to their possible ancestry has run along two main lines. They have been regarded as derived either from a phycomycetous group, or from the Red Algae or the ancestors of the latter. An independent origin either among the flagellates or the Green Algae has also been proposed.

The suggestion of a floridean relationship was first made by Sachs, in 1874, and has recently been supported by Harper, Dodge[1] and others; the

[1] *Bull. Torrey Bot. Club*, 1914, xli, p. 157.

comparison of the sporogenous filaments of a Red Alga with the ascogenous hyphae, of the algal trichogyne with the unicellular trichogynes of some Ascomycetes, and of the spermatia in the two groups is certainly suggestive, but it assumes that the trichogyne in the two cases is homologous, a very doubtful point, and it involves the corollary that all the simpler Ascomycetes are derived by reduction from the more complex.

The derivation of the group from the Phycomycetes was upheld by Winter in 1874, by de Bary in and after 1881, by Brefeld in 1889, and lately by Atkinson[1] in 1914.

Any conclusion however, as to either the origin or the inter-relationships of the Ascomycetes, must await a detailed knowledge of the development of the ascocarp and of the morphology of the sexual apparatus in a much larger number of species.

So far the development of the ascocarp throws little light on the problems of phylogeny; in the great majority of cases the ascogenous hyphae are enclosed when young by a weft of gametophytic filaments, and, in the simplest of such cases, this arrangement persists till the asci are ripe and the gametophytic hyphae decay. In other cases there is a considerable variation in the time and extent of the opening of the fructification, though its mature form is of two main types. In view of this fact it is clear that the Pyrenomycetes, in which the perithecium is flask-shaped, opening by an ostiole, and the Discomycetes, in which the apothecium is typically cup-shaped and wide open at maturity, are justified as form groups and this long established arrangement is not contradicted by what we know of the development of the sexual organs.

The **Discomycetes** show three types of sexual apparatus:

(1) The oval antheridium and somewhat elongated, coiled archicarp of *Ascodesmis*. Here the archicarp ends in a one-celled trichogyne and the unicellular oogonium becomes septate after fertilization and gives rise to a few short ascogenous hyphae[2].

(2) An oval antheridium, unicellular trichogyne, and more or less spherical oogonium which does not undergo septation. This type is found in *Pyronema*, and may be traced in *Humaria granulata*, and presumably in other forms where a large oval cell has been seen at the base of the developing ascocarp. It is present also in *Lachnea stercorea*, if the septate trichogyne of this species can be looked upon as a secondary development. It is the characteristic type of the Pezizaceae.

(3) A scolecite or stout septate archicarp, the distal cells of which form a trichogyne and the proximal a stalk, while the middle region is multicellular,

[1] *Ann. Mis. Bot. Gard.* 1914, ii, p. 315.

[2] Another simple form is *Thelebolus*, which shows certain suggestive analogies to *Sphaerotheca*, but demands further investigation.

the cells communicating by means of large pits, and one (*Ascobolus furfuraceus*) or more (*Ascophanus carneus*, *Lachnea cretea*, *Rhizina undulata*) of them give rise to ascogenous hyphae. This is the characteristic archicarp of the Ascobolaceae and is found also in *Lachnea cretea* and in *Rhizina undulata* as well as in *Collema pulposum* and some other lichens. Unfortunately the details of normal fertilization are not yet available in these forms so that it cannot be said definitely whether the oogonial region is multi- or unicellular at the fertilization stage. In many cases fertilization is known to be "reduced" and in the majority an antheridium has not been recorded. In *Ascobolus carbonarius*, a small, spermatium-like, attached cell (the male conidium of Dodge) has been observed in association with the tip of the trichogyne; somewhat similar attached cells have been seen in *Collema pulposum*; in other lichens typical, detached spermatia have been described. Here the type of maximum septation is reached and shows a septate trichogyne, an oogonial region sooner or later septate, and a spermatium-like antheridium.

The second or spherical type may be derived very readily from the first; the genera *Ascodesmis* and *Pyronema* are alike in the development of their sexual organs and the structure of their sheath; the significant difference between them lies in the spherical shape of the oogonium in *Pyronema* which may well be responsible for the absence of septation. The third or scolecitic type could also be derived through an increase of septation and reduction in the size of the antheridium from *Ascodesmis*; the coloured spores and reticulate epispore characteristic of that genus and of many Ascobolaceae may be, as Massee suggests, a further indication of relationship.

The sexual apparatus of the **Pyrenomycetes** culminates in a septate trichogyne, septate oogonial region and detached antheridium (spermatium) so closely comparable with those of the discomycetous type of maximum septation as to suggest some cross relationship between the groups. The very distinct type of ascocarp, however, in *Gnomonia*, *Polystigma* and other Pyrenomycetes on the one hand, and in the Ascobolaceae and their allies on the other appears at present to negative this possibility and to indicate rather a case of parallel development.

In the Laboulbeniales the trichogyne is septate, and the oogonium is unicellular and undergoes septation after the fertilization stage, but, of the row produced, only one cell gives rise to asci. The fertilizing agent may be a walled spermatium budded off externally, or it may be a non-motile mass of protoplasm ejected from an attached organ.

The detailed development of the lower Pyrenomycetes has as yet been even less studied than that of the higher forms, and all that can be said is that apparently the antheridium is of a simple attached type and that the archicarp is somewhat elongated, often coiled and sooner or later septate. The larger gametangia among these forms would well repay detailed investigation.

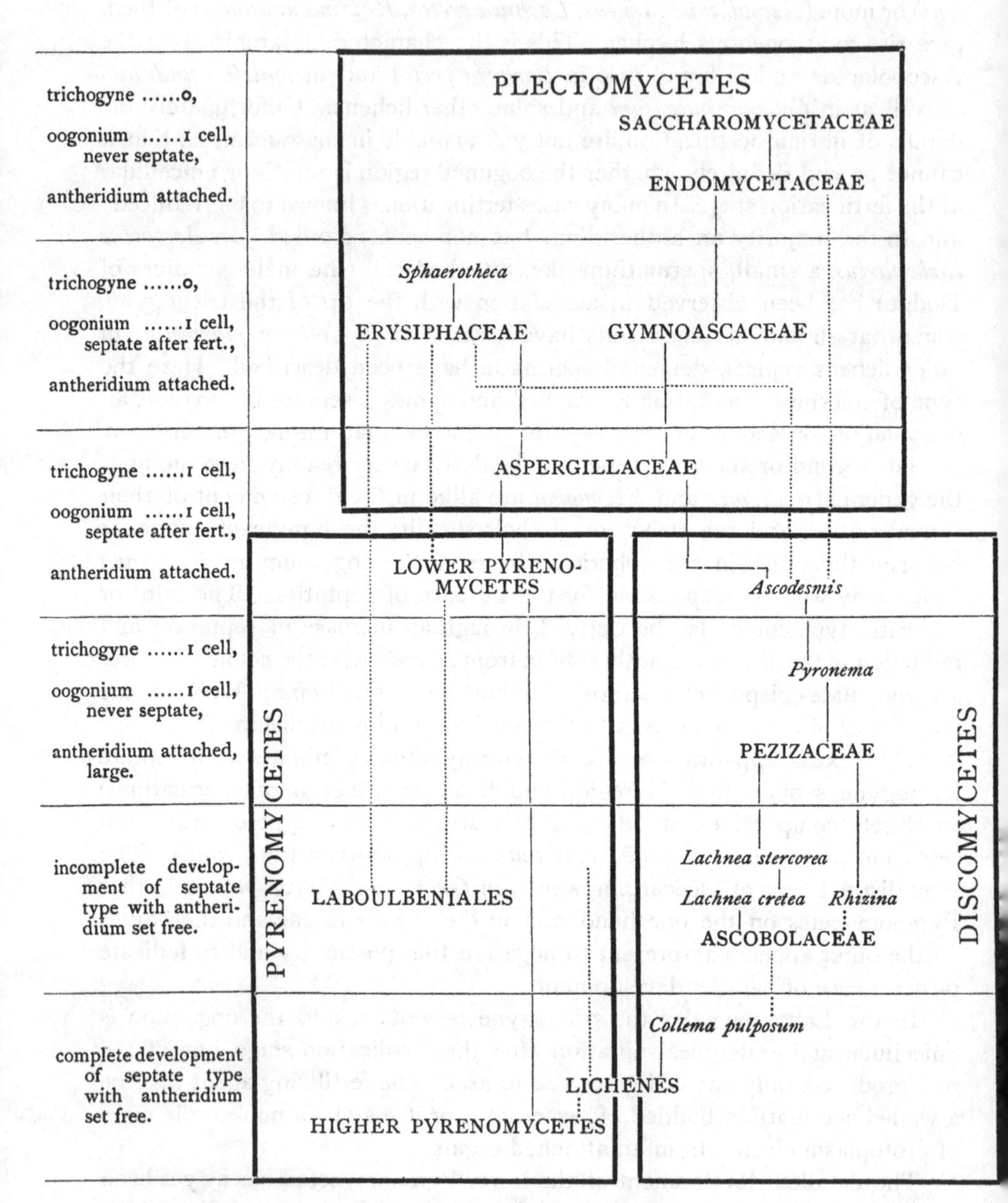


The dotted lines indicate possible and the continuous lines rather more probable relationships.

In the forms here grouped together as **Plectomycetes**, we have three or four varieties of sexual apparatus, but the richly septate types are never reached and detached spermatia are not known. The antheridium is a uninucleate or coenocytic cell borne at the end of a stalk; the most elaborate archicarp is that of *Eurotium* with a unicellular trichogyne and an oogonium which becomes septate after the fertilization stage. In *Gymnoascus* there is no trichogyne and the compact sheath of the Aspergillaceae is represented by an open weft of hyphae.

In the Erysiphaceae a trichogyne is not developed and the oogonium becomes septate after fertilization, but only one cell gives rise to ascogenous hyphae or (in *Sphaerotheca*) becomes the single ascus.

In the Endomycetaceae two enlarged cells unite, and the single ascus is the immediate product of their union. The fusing cells arise on the same mycelium as the conidia, and in *Endomyces* they are of different size; each is uninucleate and after the union of their nuclei the fusion nucleus divides to form the nuclei of the ascospores which may be four or eight in number; in view of what is known in other asci it may be inferred that in the course of these divisions meiotic reduction takes place.

Should this inference prove correct, and there seems no other stage of the life-history at which meiosis is likely to occur, we have here a diploid phase of the briefest possible duration, meiosis immediately succeeding fertilization. This condition may be either primitive or reduced; ascogenous hyphae have not yet been interpolated or have already disappeared. The only indication in favour of the latter hypothesis is the occurrence in certain species of three nuclear divisions in the ascus and eight ascospores; there are, however, other cases, such as the oogonium of the Fucaceae, where a third division regularly follows meiosis even when all the nuclei formed are not to be utilized, and in consequence the significance of the third mitosis cannot be pressed. If the Endomycetaceae be regarded as primitive, or rather as a simple offshoot from a primitive common ancestor, the ascogenous hyphae must be regarded as an interpolated phase in the life-history, and this interpolation seems to have entailed (1) the septation of the cell in which fertilization takes place, and (2) the formation of one or more asci from one or more of its subdivisions.

These changes have been established in certain simple forms through which the rest may have been derived:

(1) Erysiphaceae; the vegetative cells are generally uninucleate; the oogonium after fertilization divides to form a row of cells, and from one of these ascogenous hyphae grow out. The single ascus of *Sphaerotheca* and *Podosphaera* is presumably the result of reduction, since there seems no explanation of the development of a sporophyte unless the number of spores resulting from a single sexual act is thereby increased. As simple forms go, the sheath is

here fairly elaborate; the asci are regularly arranged. The Erysiphaceae may possibly have given rise to the Laboulbeniales, the only other group in which a single daughter cell of the oogonium is responsible for the asci, and perhaps to the lower Pyrenomycetes also; these, like the Erysiphaceae, have regularly arranged asci, and in *Chaetomium fimete* the perithecium is without an ostiole.

(2) *Gymnoascus* and its allies; the gametangia are not much differentiated; both, like the mycelial cells, are multinucleate (it is difficult to gauge the significance of this character in Fungi) and the oogonium, or an outgrowth from it, becomes septate after fertilization and gives rise to ascogenous hyphae. Among the Endomycetaceae, *Dipodascus* suggests itself as the most probable representative of the ancestor of these forms. The asci in the Gymnoascaceae are irregularly arranged and the sheath is rudimentary. Clearly such a form may have given rise to the Aspergillaceae and to the rest of the higher Plectascales but perhaps to no other group.

(3) *Ascodesmis* is a third type which might be derived either directly or through the erysiphaceous type from an endomycetous ancestor; the antheridium and oogonium are but little differentiated, but the latter is furnished with a trichogyne and becomes septate after fertilization; the ascogenous hyphae are few and the sheath simple. Massee indeed places this genus near *Gymnoascus* although the asci are regularly arranged. It cannot be far removed from the ancestor of the other Discomycetes.

The table on page 52 may perhaps serve to elucidate some of these hypotheses.

# CHAPTER III

## PLECTOMYCETES

THE group **Plectomycetes** is constituted to include those relatively simple forms which possess neither the cup-shaped apothecium of the Discomycetes, nor the flask-shaped perithecium opening by a definite ostiole which characterizes the Pyrenomycetes. In the majority of the remaining Ascomycetes a rounded ascocarp is produced, but it opens either by the decay of its walls, or by an irregular split or tear. The asci may arise from the floor of this fructification, and stand parallel one to another, or they may be irregularly disposed, the fertile hyphae forming a tangled weft. In other families the asci are naked; they stand parallel in the Exoascaceae, but in these parasitic forms their position is probably determined by the fact that they grow up between the epidermal cells or under the cuticle of the host, and may be without phylogenetic significance. In the Endomycetaceae they are irregularly disposed on the mycelium, and in the Saccharomycetaceae a mycelium is not developed.

In the majority of species the asci are club-shaped, pyriform or oval, they arise indifferently from the terminal or intercalary cells of the fertile hyphae, and the regular bending over of the tip of the ascogenous filament characteristic of the Discomycetes and Pyrenomycetes is not found among them[1]. The ascospores are usually continuous and hyaline; in the large majority of cases the gametophytic mycelium gives rise to conidia.

In the Exoascales the asci arise on a mycelium of binucleate cells and the origin of the binucleate arrangement is unknown; but in the other main groups of Plectomycetes the form of the sexual organs has been recorded, and a number of species show functional sexuality.

In the Endomycetaceae and Saccharomycetaceae fusion takes place between similar or nearly similar cells, a single fusion nucleus is formed in the zygote and there gives rise to the nuclei of the ascospores.

In the Erysiphaceae the mycelial cells and gametangia are uninucleate the antheridium and oogonium differ in size and after fertilization the zygote (oogonium) undergoes septation and one of its cells either becomes the single ascus or branches and gives rise to several asci.

In the Gymnoascaceae and Aspergillaceae the cells of the mycelium and sexual branches are multinucleate, the oogonium is furnished with a

[1] The group corresponds, therefore, to Dangeard's Rectascées with the addition of his Gametangiées and Choristogamétées (*Le Botaniste*, 1907, p. 28).

trichogyne and the zygote (oogonium) after septation gives rise to numerous asci which arise from its several cells.

In none of the Plectomycetes are the more complicated forms of sexual apparatus reached; the trichogyne and oogonial region are never multicellular and the antheridium never becomes a spermatium.

The structure of the reproductive branches thus bears out the inference drawn from the characters of the ascigerous stage that the species included under the Plectomycetes are simple and for the most part presumably primitive. The group is a collection of forms at a somewhat similar level of development, and may or may not prove to be a natural arrangement; in this it resembles the Discomycetes and Pyrenomycetes.

It is within this group that the vegetative sporophyte has been developed and the fusion in the ascus established, and it is probably among these forms that the explanation of the peculiarities of ascomycetous morphology is to be sought.

The group Plectomycetes includes some 1200 species and may be subdivided as follows:

| | |
|---|---|
| Asci irregularly arranged | PLECTASCALES. |
| Asci parallel | |
| Ascocarp developed | ERYSIPHALES. |
| Ascocarp lacking | EXOASCALES. |

## PLECTASCALES

The Plectascales include all those Ascomycetes in which the asci are arranged irregularly, at different levels and diversely orientated. In the better developed forms the ascogenous hyphae are enclosed in a definite peridium made up of an inner nutritive and an outer protective sheath, or they are surrounded, as in *Gymnoascus*, merely by an open weft of hyphae. The latter arrangement suggests a connection with the Endomycetaceae, a series of simple species in which the asci are quite unprotected; it has been shown, by the recent studies of Guilliermond and others, that these in turn lead to species in which a mycelium is seldom developed, and finally to the typical yeasts, where the endosporogenous cells are scarcely recognizable as asci.

To these may be added a few other simple forms of uncertain relationship so that the Plectascales include the Plectascineae, the Protascineae, and some of the Hemiasci of the authors of Engler's *Natürlichen Pflanzenfamilien*, and contain some of the simplest members of the Ascomycetes.

Among these the Saccharomycetaceae have presumably been derived by reduction from such forms as the Endomycetaceae where a mycelium is normally developed. In both groups a single fusion occurs in the life-history and immediately precedes the formation of the ascospores. It might be possible to regard this fusion as representing the non-sexual fusion in the ascus which characterizes the members of other groups of Ascomycetes, and to look upon the gametophyte and sexual organs as having wholly disappeared.

On the other hand, the fact that fusion is between separate cells, and not between nuclei located in the same cell, and the further fact that the fusing cells sometimes differ in size and behaviour (and not infrequently fail to fuse) recall characteristics of the sexual process in other fungi, and indicate that we are dealing here with forms in which the vegetative hyphae are gametophytic, the uniting cells are gametangia, the diplophase is represented only by the fusion nucleus, the ascospores are formed directly in the oogonium, and the anomalous second fusion does not occur. In *Guilliermondia*, where the fusion cell buds out a daughter cell in which the ascospore is produced, we may conceivably have the rudiment—or the vestige—of an ascogenous hypha, and in *Dipodascus* the outgrowth of the gametangium after fertilization may have a similar significance. Should the Endomycetaceae prove to be primitive, the history of the higher Ascomycetes becomes that of the interpolation of a vegetative sporophyte between fertilization and meiosis.

The Plectascales, as defined above, include the following families:

| | |
|---|---|
| Asci naked. | |
| Vegetative cells forming a mycelium; asci distinct from vegetative cells | ENDOMYCETACEAE. |
| Vegetative cells single or loosely attached; asci not differentiated from vegetative cells | SACCHAROMYCETACEAE. |
| Asci surrounded by loosely interwoven hyphae | GYMNOASCACEAE. |
| Asci surrounded by a definite peridium. | |
| Ascocarp subaerial | |
| Sessile | ASPERGILLACEAE. |
| Stalked | ONYGENACEAE. |
| Ascocarp subterranean | |
| Peridium distinct from walls of ascocarp; spore mass powdery at maturity | ELAPHOMYCETACEAE. |
| Peridium continuous with walls of ascocarp; spore mass never powdery | TERFEZIACEAE. |

## *Endomycetaceae*

In the Endomycetaceae the mycelium is usually well developed and bears numerous asci each of which is either the product of a separate and presumably sexual fusion, or parthenogenetically produced. Oidia, chlamydospores and yeast-like conidia may be formed. The majority of the Endomycetaceae are saprophytic on sugary substances or on exudations from plants. *Endomyces Mali* is described as an active parasite on apples and various species are parasitic on other fungi. The principal genera are *Eremascus* and *Endomyces.*

Only two species of *Eremascus* are known. *E. albus* was discovered by Eidam in 1881, in a bottle of malt extract. The contents had gone bad and their surface was covered with a growth of various fungi, amongst which was the new genus. It produced a fine, snowy white, septate mycelium from which pairs of fertile hyphae grew out, curled round one another and fused at their tips (fig. 18). The fused portion was cut off from the fertile hypha below, and eventually produced eight spores. Unfortunately Eidam's species was lost and has never reappeared.

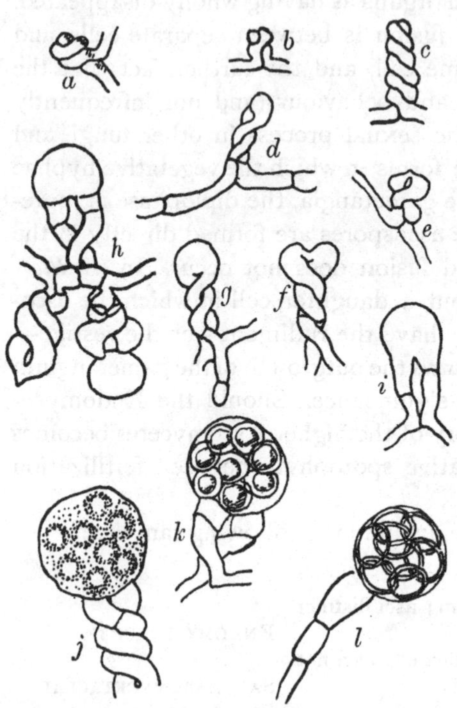


Fig. 18. *Eremascus albus* Eidam; *a. b. c. d.* sexual apparatus; *e. f. g. h.* fusion of gametangia; *i. j. k.* development of asci; *l.* parthenogenetic ascus; ×900–1000; after Eidam.

In 1907, however, Stoppel, on opening some pots of apple and gooseberry jelly, discovered a very similar form which she named *Eremascus fertilis.* This species, like *E. albus*, possesses a branching, septate mycelium. The cells at first contain several nuclei; these, according to Stoppel, are arranged in pairs, but Guilliermond, in a subsequent investigation, found that such an arrangement, even when present in the young mycelium, did not persist. It is no doubt dependent upon rapidity of growth.

From this mycelium pairs of uninucleate branches grow up, usually from the same, sometimes from different hyphae, and fuse at their apices (fig. 19). Their nuclei also fuse and after three karyokinetic divisions eight spores are formed. Sometimes, especially in old cultures, the fertile hyphae may produce asci without fusion. These are usually small and generally contain four spores or even a lesser number. As a rule three nuclear divisions take place in the parthenogenetic asci, and eight nuclei are formed, though they do not all function. According to Guilliermond it would seem that the number of spores is conditioned not by any cytological peculiarity, but rather by the supply of nutritive material.

The species of the genus *Endomyces* possess a branched, septate mycelium. It may break up into oidia, which sometimes become surrounded by thick walls and form cysts, or it may produce yeast-like conidia which

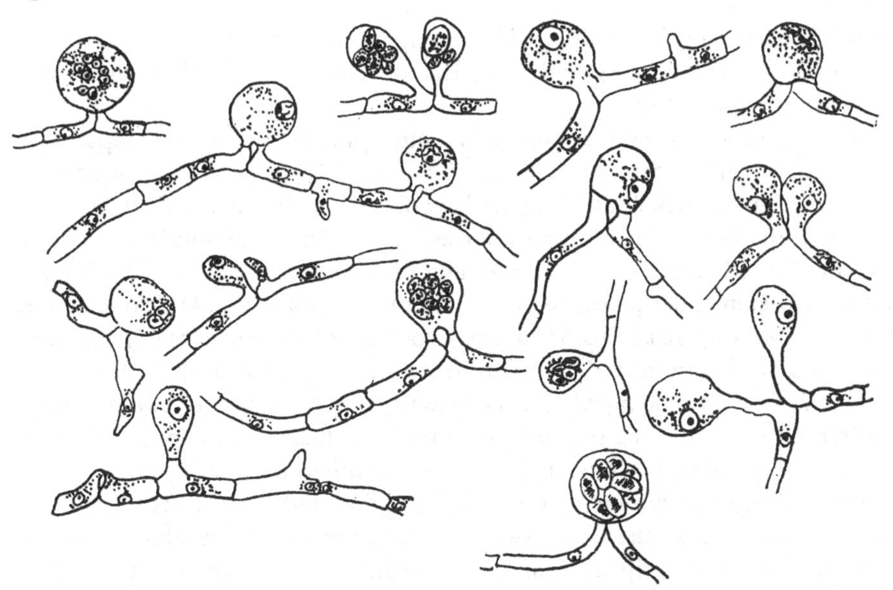

Fig. 19. *Eremascus fertilis* Stoppel; stages in the formation of the ascus, both by fusion of two cells and parthenogenetically; after Guilliermond.

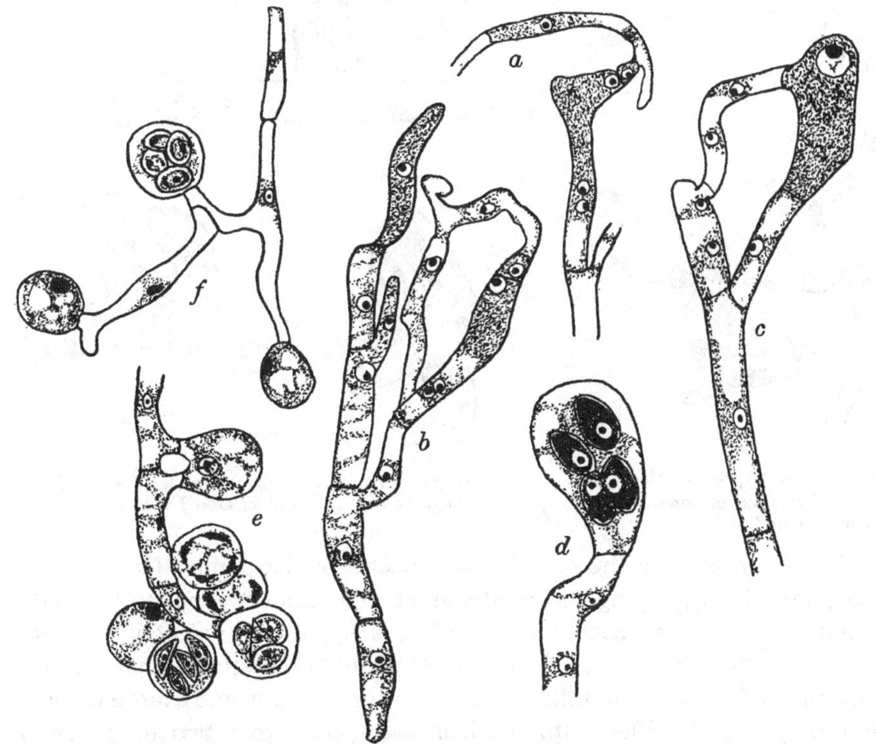


Fig. 20. *Endomyces Magnusii* Ludw.; *a. b.* antheridium and oogonium in contact; *c.* oogonium after fusion of sexual nuclei; *d.* parthenogenetic ascus; *Endomyces fibuliger* Lindner; *e.* conjugation between two neighbouring cells, at the end of the hypha is a group of young asci; *f.* normal and parthenogenetic asci; after Guilliermond.

themselves multiply by budding. Under appropriate conditions the mycelium also bears naked, four-spored asci, the development of which has been studied by Guilliermond.

In *Endomyces Magnusii* vegetative multiplication is by the separation of oidia cut off by transverse walls. The ascus is the product of a definite sexual process in which an elongated, swollen cell and a relatively narrow one grow up, bend towards one another and unite. The single nucleus of the smaller cell passes into the larger and fuses with its nucleus. Two karyokinetic divisions take place, so that four nuclei and ultimately four spores are produced (fig. 20). Fusion appears to take place indifferently between unrelated or closely related filaments and parthenogenesis is not uncommon.

In *E. fibuliger*, about half the asci result from the union of two filaments while the remainder are parthenogenetic. The fusing hyphae are in most cases closely related, often appearing as protuberances on each side of a septum. They are similar at the time of fusion, but afterwards the growth of one ceases, while the other swells to form the ascus. Besides producing asci *E. fibuliger* multiplies rapidly by means of yeast-like conidia, which closely resemble the cells of species of *Saccharomyces* (fig. 21).

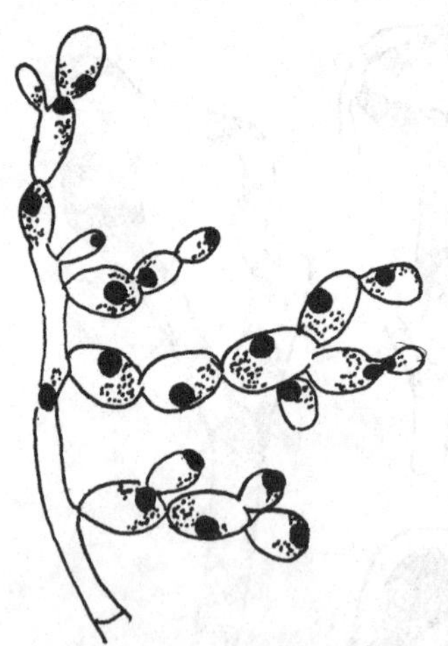

Fig. 21. *Endomyces fibuliger* Lindner; formation of conidia; after Guilliermond.

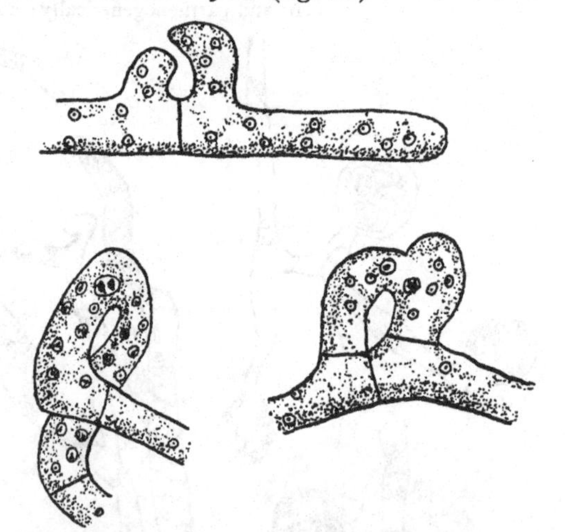

Fig. 22. *Dipodascus albidus* Lagerh.; fusion of conjugating cells and nuclei; after Dangeard.

*Wolkia decolorans*, the only known species of the genus *Wolkia*, has a strong mycelium, growing luxuriantly at a temperature of about 26° C.; in the immature state the mycelium is light pink, later globular red bodies appear. These are the asci and are formed at the ends of single hyphae without preliminary fusion. At first they contain dense cytoplasm and a large vacuole; later they become filled with blackish ascospores from two to fifteen in number; in old cultures the spores are sometimes septate.

*W. decolorans* has been identified by van der Wolk as the cause of "yellow grains," a serious disease of stored rice which is endemic in the East Indies and elsewhere, but is prevented by attention to thoroughly dry conditions.

The only known species of *Dipodascus*, *D. albidus*, was discovered by de Lagerheim in Quinto in 1892 on the gummy secretion of *Puya*, one of the Bromeliaceae. It was found again in 1901 by Juel, in Sweden, on the sap of an injured birch, and has been described by both these investigators and by Dangeard. The mycelium consists of multinucleate cells; it may break up to form oidia and may produce many-spored asci.

In the initiation of the ascus two branches grow up from the same or different hyphae and fuse at their apices (fig. 22). Both are multinucleate; one is rather larger than the other and continues its growth after fertilization to form the single ascus. Soon after the fusion of the sexual cells one of the nuclei in each is recognizable as larger than its neighbours. These two nuclei unite, usually in the larger cell, while the rest of the nuclei degenerate. The fusion nucleus passes up into the developing ascus, undergoes several divisions and eventually gives rise to the nuclei of the spores (fig. 23).

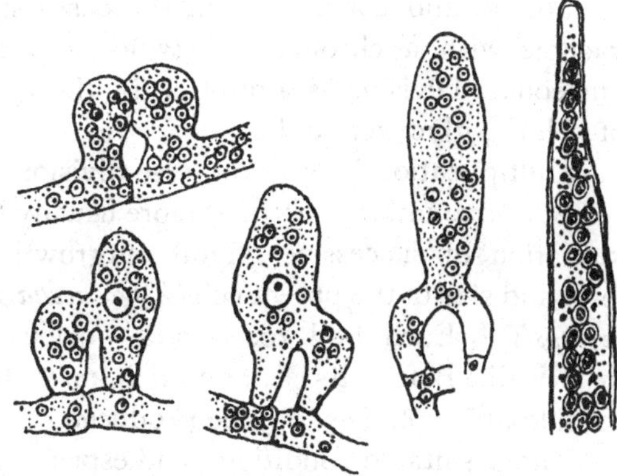

Fig. 23. *Dipodascus albidus* Lagerh.; development of ascus and ascospores; after Juel.

*Dipodascus* differs from *Eremascus* and the other Endomycetaceae in the presence of accessory nuclei in its gametangia, and from all except *Wolkia*, in the formation of numerous spores in its ascus. In spite of these differences the resemblance seems sufficiently close to permit its inclusion in the same group.

## ENDOMYCETACEAE: BIBLIOGRAPHY

1883 EIDAM, E. Zur Kenntniss der Entwickelung bei den Ascomyceten. Cohn's Beiträge zur Biol. der Pflanzen, iii, p. 385.

1892 DE LAGERHEIM, G. *Dipodascus albidus* eine neue geschlechtliche Hemiascee. Jahrb. für. wiss. Bot. xxiv, p. 549

1902 JUEL, H. O. Über Zellinhalt, Befruchtung und Sporenbildung bei *Dipodascus*. Flora, xci, p. 47.

1907 DANGEARD, P. A. Recherches sur le développement du périthèce chez les Ascomycètes. Le Botaniste, x, p. 30.

1907 STOPPEL, R. *Eremascus fertilis*, nov. spec. Flora, xcvii, p. 332.

1909 GUILLIERMOND, A. Recherches cytologiques et taxonomiques sur les Endomycetacées. Rev. Gén. de Bot. xxi, p. 1.

1909 KLÖCKER, A. *Endomyces Javanensis*, n. sp. C. R. des trav. du lab. de Carlsberg, vii, p. 267.
1913 VAN DER WOLK, P. C. *Protascus colorans* a new genus and a new species of the Protascineae Group; the source of "Yellow-Grains" in Rice. Myc. Centralbl. iii, p. 153.
1914 RAMSBOTTOM, J. The Generic Name *Protascus*. Trans. Brit. Myc. Soc. v, p. 143.

### *Saccharomycetaceae*

The Saccharomycetaceae, or yeasts, are widely distributed on, or in, all sorts of sugary media; they occur mainly as separate cells which are only exceptionally united to form a short mycelium.

The individual cells are round or elliptical, bounded by a delicate membrane and containing, in the cases studied in detail, a large nuclear vacuole with a chromatin network and a well-marked, laterally placed nucleolus. Division is amitotic. In the cytoplasm are refractive granules of volutin, glycogen and oil.

Multiplication is by transverse division and separation of the daughter cells (*Schizosaccharomyces*), or more usually by budding, that is to say by the formation of successive lateral outgrowths which ultimately assume the form and size of the parent cell (*Zygosaccharomyces*, *Saccharomyces*, *Saccharomycopsis*). Each bud receives a nucleus and cytoplasm and is cut off by a wall. Before its separation it may itself bud again, and in this way considerable colonies may be produced.

Under suitable conditions, and especially when growing on a moist, solid substratum, the cell contents may round themselves up to form one to eight (usually two or four) spores. These so-called **endospores** are the ascospores of the yeast, the ordinary vegetative cell functioning as an ascus either independently or after conjugation with another similar cell.

One of the most striking features of the yeasts and one which gives them a considerable economic importance is the power possessed by many species of producing alcoholic fermentation in certain sugars. This property is due to the presence of the enzyme **zymase** which is secreted by the yeast cells during fermentation, but which is not present in the resting cells, being soon decomposed when the reaction comes to an end. The activity of zymase is dependent on the presence of two co-enzymes; the first is a soluble phosphate which enters into temporary combination with part of the carbohydrate, but it is ineffective in the absence of a second factor of unknown constitution. The unknown co-enzyme is dialysable and not destroyed by boiling; it may be separated from the yeast juice by filtration under pressure, both filtrate and residue being inactive alone.

Even the yeasts which produce the largest proportion of alcohol utilize five to six per cent. of the available sugar in the formation of glycerine, succinic acid, acetic acid, and small quantities of other substances. The amount of these by-products varies during the progress of fermentation and according to external conditions. In particular, fermentation is affected by the presence or absence of free oxygen. Under conditions of plentiful aeration the yeast grows and multiplies rapidly and much of the sugar is used as food; under anaerobic conditions, on the other hand, the main part of the sugar is utilized in respiration, alcoholic fermentation is more complete and the quantity of alcohol produced is greater in proportion to the number of cells concerned.

On the ground that their daughter cells are produced by septation, and not, as in other genera, by budding, Guilliermond postulates for the species of *Schizosaccharomyces* a derivation from the neighbourhood of *Endomyces Magnusii* in which the mycelium cuts off free cells by transverse septation. He refers such genera as *Saccharomyces*, in which budding occurs, to the line which gave rise to *Endomyces fibuliger* where asexual multiplication takes place in a similar way (fig. 24).

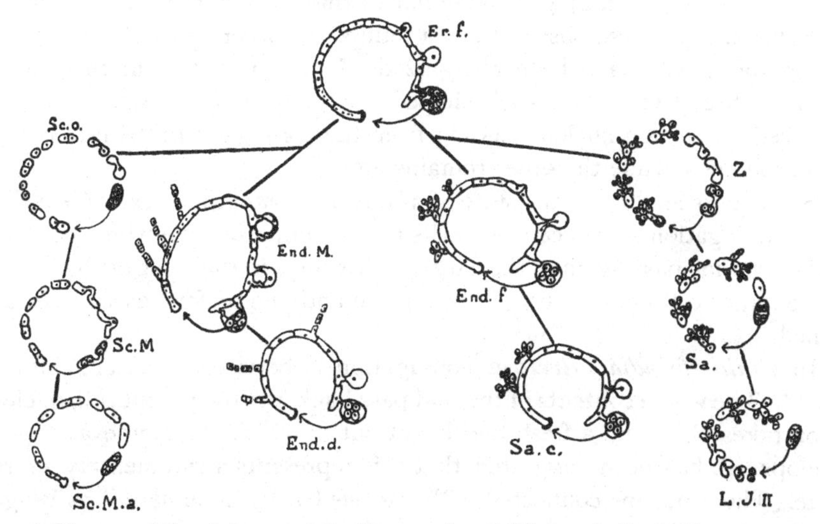


Fig. 24. Diagram of the phylogeny of the Yeasts; after Guilliermond. Er. f., *Eremascus fertilis.* End. f., *Endomyces fibuliger.* Sa. c., *Saccharomycopsis capsularis.* Z., *Zygosaccharomyces.* Sa., *Saccharomyces.* L. J. II, Johannesberg yeast II. End. M., *Endomyces Magnusii.* End. d., *Endomyces decipiens.* Sc. o., *Schizosaccharomyces octosporus.* Sc. M., *Schizosaccharomyces mellacei.* Sc. M. a., *Sch. mellacei*, apogamous variety.

Conjugation, as a preliminary to the formation of asci, was first described by Schiönning in 1895 in *Schizosaccharomyces* and was afterwards studied in some cytological detail by Guilliermond (1901). In *Sch. octosporus*, two neighbouring cells of similar size put out processes which fuse to form a conjugation tube; the nuclei pass into the tube and undergo fusion, after which the two associated cells enlarge and form, as a rule, a single oval structure, the ascus. The fusion nucleus divides to form four or eight daughter nuclei about which ascospores are organized (fig. 25). Sometimes the limits of the conjugating cells can be distinguished after the ascospores are formed, two or four lying in each of the original cells.

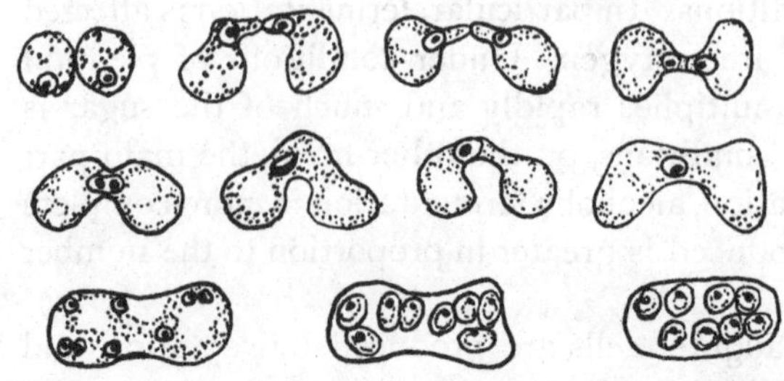

Fig. 25. *Schizosaccharomyces octosporus* Beyrinck; conjugation and formation of ascospores; after Guilliermond.

In the closely related *Sch. Pombe* and *Sch. mellacei* copulation takes place in a very similar way, but the union of the conjugating cells is less complete than is usually the case in *Sch. octosporus*, and the number of ascospores is regularly four. The mature ascus is thus dumb-bell shaped, with two spores in each enlargement.

Also in 1901 Barker discovered the yeast *Zygosaccharomyces Barkeri* and observed in it a conjugation similar to that of *Sch. Pombe.*

Following on these observations a number of other cases of conjugation among the yeasts have been recognized. Many species form two spores, one at each end of a dumb-bell shaped ascus; in others a single spore is produced, the fusion nucleus passing from the conjugation tube into one of the fusion cells, while the other remains empty.

Such cases lead up to the state of affairs in *Zygosaccharomyces Chevalieri* where conjugation is between two cells of different sizes, the whole contents of the smaller passing into the larger cell which is then cut off by a wall. In the larger cell nuclear fusion takes place and one to four ascospores are formed.

In *Guilliermondia fulvescens* conjugation is between a mature cell and its bud. The whole contents of the bud pass back into the parent cell, nuclear fusion takes place and a fresh bud is put out in which the single ascospore develops. It has been suggested that this represents a rudimentary sporophyte. The problems connected with meiosis hardly arise here if, as Wager has shown for *Saccharomyces*, amitosis is the rule.

In *Debaryomyces globosus*, copulation takes place sometimes between two similar cells, sometimes, as in *Guilliermondia*, between a mature cell and its bud. In either case one or two spores are produced

In most species with sexually formed asci parthenogenesis is not uncommon. In *Schwanniomyces occidentalis* and in *Torulaspora Rosei* it has become the rule. The cells in which the ascospores are about to be formed put out processes which are directed towards neighbouring cells at the same stage of development. But they do not fuse, and ascospores are formed in each cell independently. In *T. Rosei* more than one process is sometimes put out.

Such cases lead up to the conditions found in the large majority of yeasts where ascospore formation takes place not only without effective conjugation but without any vestiges of that process.

Among forms with parthenogenetically produced asci Guilliermond (1902) has observed a peculiar process in *Saccharomycodes Ludwigii.* Here the rather elongated ascus gives rise to four ascospores; these on germination swell up, put out beaks towards each other and fuse in pairs. Fusion is usually between spores in the same ascus, but occasionally, when one of the four ascospores has degenerated, between spores of different groups. Each spore is uninucleate, the nuclei pass into the conjugating tube and there fuse. A similar process takes place in *Willia Saturnus* and in the so-called Yeast of Johannesberg, II.

Afterwards an outgrowth from the middle of the conjugation tube is developed; it may give rise to a short hypha which soon breaks up into separate cells, it may produce lateral buds in an ordinary yeast-like manner, or it may give rise at once, especially on carrot and under conditions favourable to ascus formation, to four endospores. In the last case the life-history differs from that of *Zygosaccharomyces* principally in the abbreviation of the vegetative phase, in other cases a vegetative phase is inserted not as usual before conjugation, but between conjugation and the development of the spores. Guilliermond regards the pairing of the ascospores as a secondary process following the establishment of parthenogenesis. In any case it admirably illustrates the plasticity of the simpler fungi.

## SACCHAROMYCETACEAE: BIBLIOGRAPHY

1895 SCHIÖNNING, H. Nouvelle et singulière formation de l'ascus dans une levure. Meddelelser fra Carlsberg Laboratoriet. iv, p. 30.

1901 BARKER, T. P. A Conjugating "Yeast." Proc. Roy. Soc. lxviii, p. 345.

1901 GUILLIERMOND, A. Recherches sur la sporulation des Schizosaccharomycètes. Comptes Rendus Ac. Sci. cxxxiii, p. 242.

1901 HANSEN, E. C. Grundlinien zur Systematik der Saccharomyceten. Centralbl. für Bakt. Abt. ii; xii, p. 529.

1903 GUILLIERMOND, A. Recherches sur la germination des spores chez le *Schizosaccharomyces Ludwigii.* Bull. Soc. Myc. de France, xix, p. 18.

1903 LEPESCHKIN, W. W. Zur Kenntniss der Erblichkeit bei den einzelligen Organismen. Centr. f. Bakt. Abt. ii; x, p. 145.

1905 GUILLIERMOND, A. Recherches sur la germination des spores et la conjugaison chez lès levures. Rev. Gén. de Bot. xvii, p. 337.

1909 KLÖCKER, A. Deux nouveaux Genres de la Famille des Saccharomycètes. C. R. des trav. du lab. de Carlsberg, vii, p. 273.

1910 GUILLIERMOND, A. Sur un curieux exemple de parthénogénèse observé dans une levure. C. R. Soc. Biol. de Paris, lxviii, p. 363.

1910 GUILLIERMOND, A. Quelques remarques sur la copulation des levures. Ann. Myc. vii, p. 287.

1910 WAGER, H. and PENISTON, A. Cytological Observations on the Yeast Plant. Ann. Bot. xxiv, p. 45.
1911 GUILLIERMOND, A. Les Progrès de la Cytologie des Champignons. Prog. Rei Bot. iv, pp. 433 and 468, *et seq.*
1911 NADSON, C. A. and KONOKOTINE, A. G. *Guilliermondia*, un nouveau genre de la famille des Saccharomycetes à copulation hétérogamique. Bull. du Jard. Imp. de St Pétersbourg, xi, p. 117.
1913 MARCHAND, H. La conjugaison des spores chez les levures. Rev. Gén. de Bot. xxv, p. 207.
1914 BAYLISS, W. M. The Nature of Enzyme Action. Monographs on Biochemistry. Longmans, Green & Co., London (and see literature cited).

*Gymnoascaceae*

The Gymnoascaceae differ from the Endomycetaceae in that their asci are borne on a sporophytic mycelium which originates from the female organ after the fertilization stage. These ascogenous hyphae are surrounded by a loose weft of protective filaments which bear spines or variously coiled or hooked branches (fig. 26). The asci are ovoid or pyriform, and each contains eight spores.

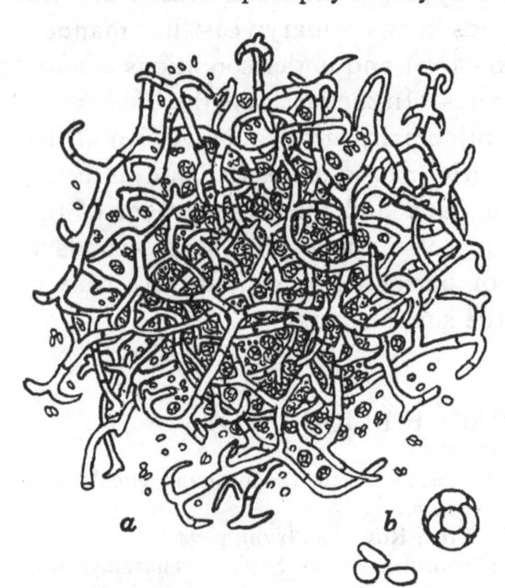


Fig. 26. *Gymnoascus* sp.; *a.* ascocarp, × 265; *b.* ascus and free ascospores, × 1040.

The species of *Gymnoascus* occur in various habitats, on dung, bees' nests, dead grass, etc.

In *G. Reesii*, according to Dale, two branches grow up from the same hypha, one on each side of a septum, and become twisted around one another. These are the antheridium and oogonium; their free ends swell into club-shaped heads which lie in close contact and each becomes delimited by a transverse septum. The walls between them break down, and the contents of the antheridium pass over into the oogonium (fig. 27 *a*, *b*). Both cells are at first uninucleate, but later coenocytic, and, though the history of the nuclei has not been traced, it seems almost certain that fusion must sooner or later occur.

Up to this point the sexual cells are usually quite similar in form, but now the antheridium grows larger and more spherical, remaining almost straight, while the oogonium puts out a prolongation which winds around it, undergoes septation and soon branches to give rise to ascogenous hyphae (fig. 27 *c*).

In *G. candidus* (fig. 27 *d*) the antheridium and oogonium already differ in form at the time of their union, and, in the majority of cases, appear to

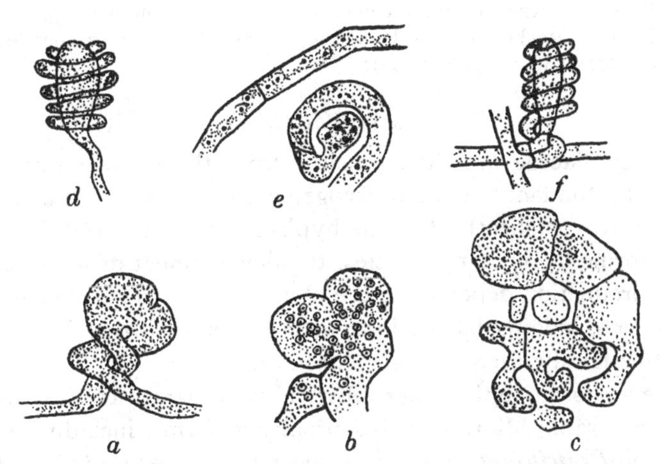


Fig. 27. *Gymnoascus Reesii* Baran.; *a.* surface view of conjugating cells; *b.* the same in longitudinal section; *c.* a later stage, septate oogonium giving rise to ascogenous hyphae; *Gymnoascus candidus* Eidam; *d.* surface view of conjugating cells; *e.* same in longitudinal section; all after Dale. *Ctenomyces serratus* Eidam; *f.* surface view of conjugating cells, × 400; after Eidam.

arise from different hyphae. As in *G. Reesii* the antheridium is straight, but the oogonium elongates before fusion and grows spirally around it till the apices meet and fuse (fig. 27 *e*). Afterwards the oogonium undergoes septation and gives rise to ascogenous hyphae. The sheath of protective hyphae is very scanty, represented by a few thin-walled filaments.

*Amauroascus verrucosus* forms sexual organs which, in their early stages, closely resemble those of *G. Reesii.* Two similar multinucleate hyphae grow up, and later one of them, the oogonium, enlarges considerably; it branches without septation, and gives rise to ascogenous hyphae which are cut off by transverse walls. Our knowledge of development in this species is due to Dangeard, who has not observed fusion between the sexual cells.

In *Ctenomyces serratus* (fig. 27 *f*), which occurs saprophytically on feathers, organs quite similar to those of *G. candidus* have been described by Eidam, and more recently by Dangeard. They are multinucleate, and the oogonium is long and elaborately coiled and ultimately becomes segmented. In this species, as in the others he has studied, Dangeard regards the central hypha not as an antheridium, but as a nutritive structure which he terms a trophogone. He denies the passage of its contents into the female organ.

## GYMNOASCACEAE: BIBLIOGRAPHY

1880 EIDAM, E. Beitrag zur Kenntniss der Gymnoasceen. Cohn's Beiträge, iii, p. 267.
1903 DALE, E. Observations on the Gymnoascaceae. Ann. Bot. xvii, p. 571.
1907 DANGEARD, P. Recherches sur le développement du périthèce chez les Ascomycètes. Le Botaniste, x, pp. 86 and 97.

### *Aspergillaceae*

The Aspergillaceae are distinguished from the earlier families of the Plectascales by the fact that their ascogenous filaments are surrounded by a closely interwoven sheath of sterile hyphae so that a closed fruit or perithecium is formed. In many species, the development of an ascus-fruit is rare, and reproduction depends on the abundant and very efficient conidia; in others, which, judging by the form of their conidial fructification, should belong to this family, ascocarps are unknown.

The species occur either saprophytically or parasitically upon a wide variety of substrata. Many of the saprophytic forms, including species of *Eurotium* and *Penicillium*, grow with especial readiness on fatty substances, *Emericella erythrospora* occurs on olives, and *Monascus heterosporus* on glycerine or tallow.

Several species of *Penicillium* together with other fungi and bacteria are concerned in the "ripening" of cheese, on the proteids of which they act by means of a proteolytic enzyme.

The species of *Microascus*, *Aphanoascus* and some others are coprophilous, and several members of the family occur on wood.

The parasitic forms are less numerous; various species of *Penicillium* and *Eurotium* are pathogenic on man and animals, and some, if they obtain entrance through a wound or other aperture, are the cause of ripe rot in fruit.

*Thielavia basicola*, the only member of the family which causes an important plant disease, is sometimes separated from the Aspergillaceae and classed with the Perisporiaceae. This species infects the roots of tobacco and certain other plants, and in the early stages multiplies by means of hyaline conidia (endospores) produced inside mycelial branches, from the ends of which they are afterwards expelled, giving the infected root a mildewed appearance. Later, thick-walled brown chlamydospores are differentiated in rows at the ends of hyaline filaments so that the root is covered with a dark coating. Normal development of the root is prevented and the host killed or stunted; if recovery does not take place perithecia are developed on the dead plant.

The formation of the perithecium in most investigated species is initiated by the appearance of sexual organs, from the stalks of which the cells of the sheath arise. In *Microascus* and in *Emericella* the sheath opens by a pore, but in the majority of cases it remains closed, and the ascospores are finally liberated by its decay. The asci are spherical or pyriform and contain two to eight continuous spores, the walls of which may be variously ornamented. In both *Penicillium* and *Eurotium* the perithecium may develop an exceptionally thick wall, and pass into a resting stage sometimes several weeks in duration. Such a structure is described as a sclerotium.

In *Eurotium herbariorum*[1] the development of perithecia is readily induced by cultivation on prune agar[2] made up with forty per cent. of cane-sugar. Ripe perithecia appear after ten days or a fortnight at 15° C. A similar method should probably prove successful with other species.

The only known species of *Aphanoascus*, *A. cinnabarinus*, was found by Zukal on the dung of alligators. The character of its perithecium wall (fig. 28) suggests a transition stage between the *Gymnoascaceae*, in which it is

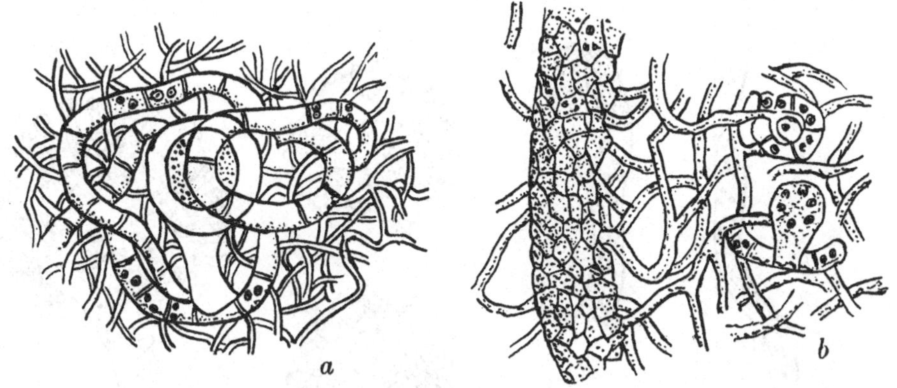


Fig. 28. *Aphanoascus cinnabarinus* Zukal; *a*. elongated, septate archicarp and swollen antheridium; *b*. ascogenous hyphae and sheath; after Dangeard.

placed by Dangeard, and the Aspergillaceae. Multinucleate conidia are borne laterally or terminally on the mycelium. The sexual organs arise from the same or different hyphae; the antheridium (Dangeard's trophogone) becomes swollen, and the elongated archicarp, which appears at first to be without septa, coils round it; their fusion has not been observed. Each at first contains from three to ten nuclei; later the number rises to about twenty in the antheridium and as many as forty in the archicarp. According to Dangeard the contents of the antheridium degenerate, though the cell itself persists for a considerable time. The archicarp undergoes septation, branches, and gives rise to asci. In the meantime vegetative hyphae grow up and invest the essential organs, forming a loose tangle around them The inner part of the investment remains in this state and is eventually absorbed, but

[1] *Eurotium herbariorum* (Wigg.) Link = *E. Aspergillus glaucus* de Bary, and *Aspergillus herbariorum* Wigg.

[2] Ten prunes are placed in a small saucepan of water and allowed to simmer, being broken open when soft; when the fluid is reduced to about 100 c.c. it is poured off, the sugar dissolved in it and five per cent. agar agar stirred in. Material grown on this medium is excellent for class work; it should be examined under the microscope while still attached to a thin slice of agar.

the outer filaments form four or five parenchymatous layers which constitute a protective sheath, apparently differing but little from that of *Eurotium* or *Penicillium*.

In the investigated species of the genus *Eurotium* (*Aspergillus*), the ascospores and conidia are commonly multinucleate and give rise on germination to a septate mycelium each cell of which contains several nuclei.

Conidiophores appear early; they arise as a rule from densely tangled knots of swollen mycelium, and appear as thick, multinucleate hyphae. The tip of the conidiophore becomes swollen and rounded off and its cytoplasm shows a reticulate arrangement. A little later numerous sterigmata bud out (fig. 29), and the nuclei stream up the strands of cytoplasm towards

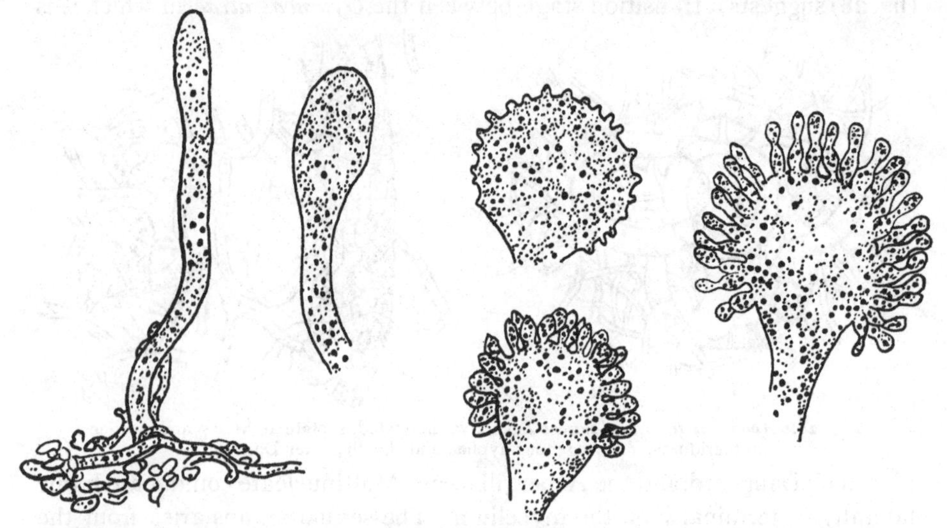

Fig. 29. *Eurotium herbariorum* (Wigg.) Link; development of conidiophores and conidia, ×625.

them. Several nuclei pass to each sterigma and thence to the conidia which develop in acropetal succession. At maturity each conidium contains in *E. repens* about twelve nuclei, in *E. herbariorum* four, and in *E. fumigatus*, *E. flavus* and *E. clavatus*, as described by Dangeard, only one.

The general features of the sexual organs of *Eurotium herbariorum* were described by de Bary in his classical researches of 1870. He distinguished a coiled, septate archicarp, and saw that its tip fused with a comparatively straight antheridial hypha, the membranes between breaking down, and he recognized that from it the asci are ultimately derived.

More recently it has been shown that the archicarp of *Eurotium* is made up of three parts: a multicellular stalk, a unicellular oogonium, and a unicellular trichogyne. In *E. herbariorum* these parts may be clearly distinguished (fig. 30 *a*), but they are not always equally definite in *E. repens*.

In any case the archicarp becomes more or less twisted and near it another septate hypha appears, from the end of which the unicellular antheridium is cut off. Like the cells of the vegetative mycelium, all parts of the sexual organs are coenocytic.

Fusion takes place between the antheridium and trichogyne, but the contents of the male organ have not been seen to enter the oogonium.

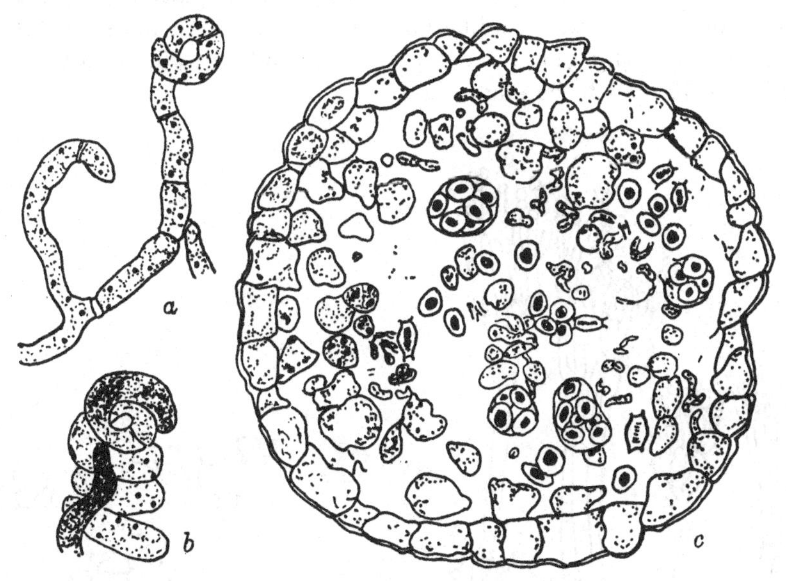


Fig. 30. *Eurotium herbariorum* (Wigg.) Link; *a.* young archicarp; *b.* archicarp and abortive antheridium; *c.* ascocarp containing asci and spores; ×625.

Dale has observed the fusion of nuclei in pairs in the oogonium of *E. repens*, and Domaradsky in that of *E. Fischeri*, but it seems probable that we are dealing here with a union of female nuclei such as occurs in various Discomycetes. Even if normal fertilization sometimes takes place it is certainly not general, for the antheridium often fails to reach the trichogyne (fig. 30 *b*), and is sometimes entirely absent. This appears to be always the case in *E. flavus* and it is common in other genera.

Nevertheless the oogonium becomes septate and from its several parts branches develop, the nuclei pass into them, and at their ends eight-spored asci develop. The ascus is formed from the terminal or the penultimate cell of a hypha and in it a fusion of two nuclei takes place.

Shortly before the septation of the oogonium, vegetative hyphae begin to grow up about the sexual organs, from the stalks of which they mostly arise as branches, and themselves branch freely. Some of the branches grow inwards forming a nutritive layer, the cells of others, as shown by de Bary, become tabular and more or less empty, they secrete a golden yellow substance, readily soluble in alcohol, in the form of a thick, brittle pellicle. These constitute the protective sheath (fig. 30 *c*), by the decay of which, after the disappearance of the nutritive hyphae and later of the ascus walls, the spores are finally set free. The ascospores are spherical or lenticular often with a sculptured epispore.

In most species of *Penicillium* reproduction takes place almost entirely by means of the abundant conidia borne in chains on the branched, brush-like conidiophores (fig. 31). Ascocarps are rare, and a detailed study of their development has yet to be made.

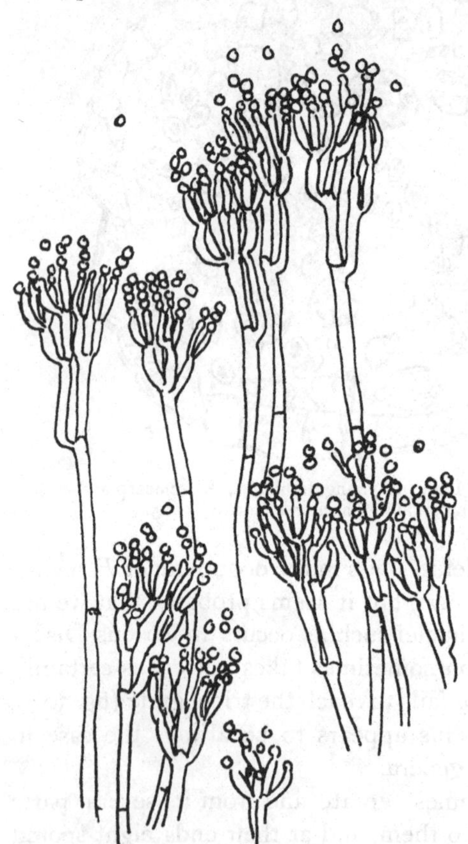

Fig. 31. *Penicillium glaucum* Link; conidiophores and conidia, × 500.

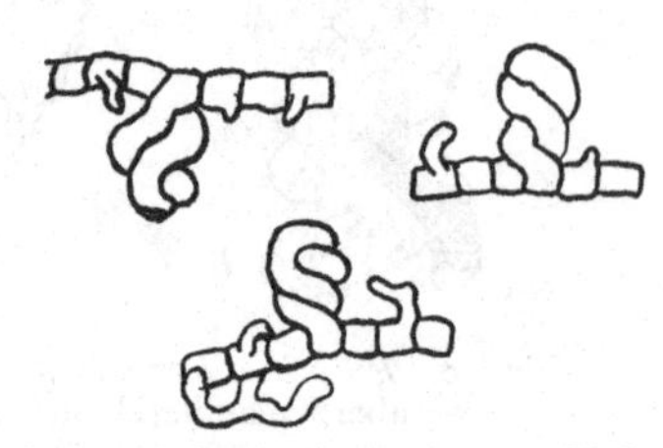

Fig. 32. *Penicillium glaucum* Link; conjugating cells, × 630; after Brefeld.

Brefeld, in 1871, succeeded in obtaining perithecia of *P. glaucum*[1] by cultivating his material under reduced oxygen pressure. He cautiously moistened slices of bread with distilled water, and after six or seven days, when the development of the mycelium was proceeding energetically,

[1] *Penicillium glaucum* Link = *P. crustaceum* L.

enclosed them between glass plates so as to reduce the entrance of air, and the development of conidia. Similarly Zukal a few years later obtained sclerotia by excluding air. The formation of the perithecia in Brefeld's material was initiated by the appearance of pairs of simple, stout hyphae which twisted round one another (fig. 32), and from one or both of which branches later arose. Brefeld regarded them as possibly oogonial and antheridial.

A further study of these organs, the simple form of which suggests a comparison with *Gymnoascus*, is much to be desired. *Penicillium glaucum* includes several biological species or strains, and it is quite possible that Brefeld's success depended not only on the methods employed, but also on the use of a fortunate variety.

Klöcker, in 1903, obtained asci in his new species *P. Wortmanni*, and another new and very curious ascigerous species, *Penicillium vermiculatum*, was described by Dangeard in 1907. The vegetative cells, unlike those of related forms, are uninucleate and bear a very scanty supply of conidia. Perithecia are abundant; in their initiation two branches take part. The oogonium is at first uninucleate but as it elongates the nucleus undergoes several divisions. In the meantime a second branch appears, usually borne on a narrower filament; it cuts off a uninucleate or occasionally binucleate terminal cell which applies itself to the middle of the oogonium, and the intervening walls disappear (fig. 33). Apparently, however, fertilization does not take place; the nucleus of the terminal cell is described as degenerating *in situ* while the oogonium undergoes septation and is transformed into a row of usually binucleate cells. Narrow vegetative hyphae grow up around this structure, and the perithecium is formed in the usual way.

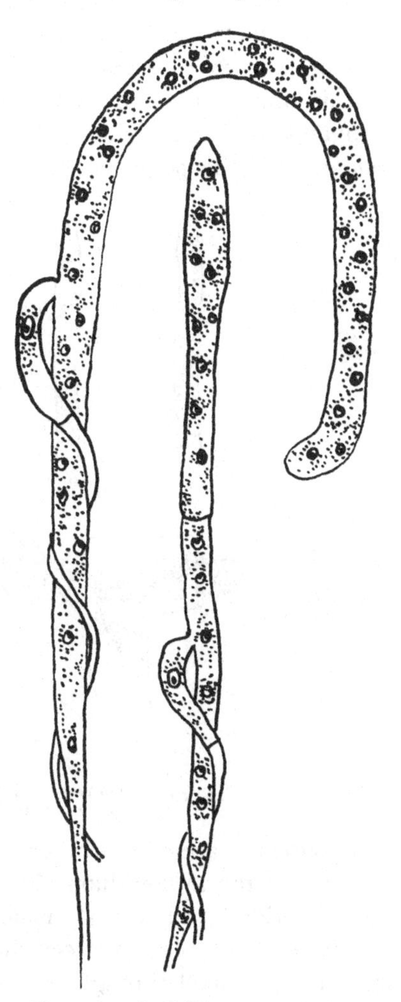

Fig. 33. *Penicillium vermiculatum* Dang.; archicarp and antheridium; after Dangeard.

The archicarp of this species shows no important peculiarities but the terminal uninucleate cell of the other branch is difficult to place; it seems almost inevitable to homologize it with an antheridium, but the relation of a uninucleate antheridium to a multinucleate female organ is by no means clear. Degeneration of the superfluous nuclei as in *Dipodascus* might be postulated. For Dangeard, the structure is a trophogone concerned only in nutrition and without sexual significance.

The genus *Monascus* contains some five or six species, of which the most fully investigated are *M. purpureus* and the form studied by Barker and since named *M. Barkeri* by Dangeard. This species is used by the Chinese for the manufacture of an alcoholic liquor known as Samsu.

In both species there is an abundant mycelium bearing chains of ovoid conidia; later it assumes a reddish tinge and fruits are formed; their development may be traced with some readiness in hanging drop cultures. Certain branches of the mycelium cut off each a small, terminal cell which elongates and bends sideways to form the antheridium (fig. 34). A prolongation of the penultimate cell grows up alongside it and a trichogyne

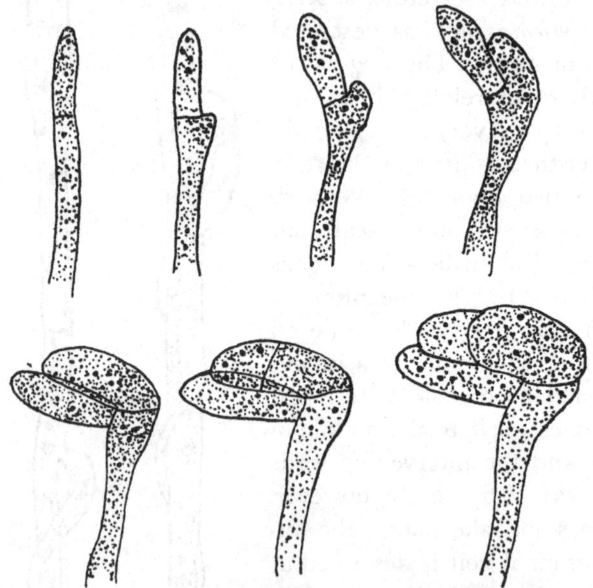

Fig. 34. *Monascus Barkeri* Dang.; development of oogonium, trichogyne and antheridium, ×900; after Barker.

and oogonium are cut off by transverse walls. The oogonium contains four to six, and the antheridium three or four nuclei. According to Dangeard the antheridial (trophogone) nuclei degenerate *in situ* but other authors find that fusion takes place between the antheridium and trichogyne and that the male nuclei travel through the trichogyne to the oogonium (fig. 35) where they pair with the female nuclei. According to Schikorra, nuclear fusion does not occur at this stage in *M. purpureus* but is postponed till the sexual nuclei have travelled in pairs to the asci, where they unite; but Kuyper, in *M. Barkeri*, reports fertilization in the female organ.

After the fertilization stage the oogonium enlarges and gives rise to ascogenous hyphae while the antheridium and trichogyne degenerate. Investing filaments grow up to form a sheath, the inner layers of which consist of delicate, nutritive cells. These degenerate early, producing a mass of protoplasm amongst which the ascogenous hyphae ramify. From the penultimate cells of the latter binucleate asci are developed, and after the nuclei have fused eight spores are formed. The ascus wall breaks down and the spores are finally set free after the decay of the outer layer of the sheath.

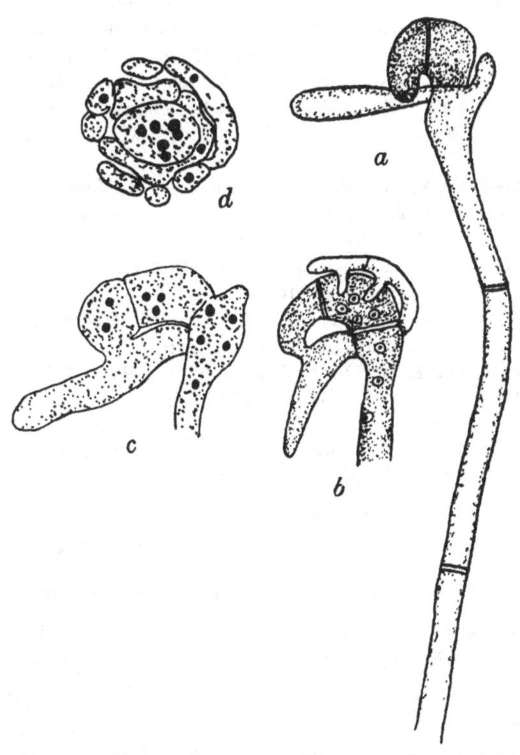


Fig. 35. *Monascus purpureus* Went.; *a. b.* stages in the development of the oogonium; after Dangeard. *Monascus X.* Schikorra; *c.* entrance of male nuclei into trichogyne; *d.* pairing of nuclei in the oogonium; after Schikorra.

This sheath, with the enclosed mass of free ascospores, was long regarded as a single organ containing an indefinite number of spores; for this reason the fungus was placed in the Hemiasci and given the generic name of *Monascus*. The later stages of development are in fact difficult to follow and have been variously interpreted by different authors. Thus Dangeard describes the oogonium as undergoing septation before it branches, while Barker interprets the mass of protoplasm in which the young asci are found as the remains of the oogonium invaginated by the growth into it of the ascogenous hyphae. Kuyper and Ikeno, on similar grounds, believed the asci to be produced by free cell formation within the oogonium, no ascogenous hyphae being developed.

## ASPERGILLACEAE: BIBLIOGRAPHY

1870 DE BARY, A. *Eurotium*, *Erysiphe*, *Cincinnobolus* nebst Bemerkungen über die Geschlechtsorgane der Ascomyceten. Beitr. z. Morph. und Phys. der Pilze, iii, p. 1.

1874 BREFELD, O. Botanische Untersuchungen über Schimmelpilze, ii, *Penicillium*, p. 1.

1887 ZUKAL, H. Über Kultur der Askenfrucht von *Penicillium crustaceum*. K. K. zoo. bot. Gesells. in Wien, xxx, vii, p. 66.
1903 BARKER, B. T. P. The Morphology and Development of the Ascus in *Monascus*. Ann. Bot. xvii, p. 167.
1903 IKENO, S. Über die Sporenbildung und systematische Stellung von *Monascus purpureus* Went. Ber. d. deutsch. Bot. Gesells. xxi, p. 259.
1903 KLÖCKER, A. Om Slogtem *Penicilliums* Plads i Systemet og Beskrivelse af en ny ascusdannende Art. C. R. des trav. du lab. de Carlsberg, vi, p. 84.
1905 KUYPER, H. P. Die Perithecienentwickelung von *Monascus purpureus* Went. und *Monascus Barkeri* Dangeard, sowie die systematische Stellung dieser Pilze. Ann. Myc. iii, p. 32.
1905 OLIVE, E. W. The Morphology of *Monascus purpureus*. Bot. Gaz. xxxix, p. 56.
1907 DANGEARD, P. Recherches sur le développement du périthèce chez les Ascomycètes. Le Botaniste, x, p. 118.
1907 FRASER, H. C. I. and CHAMBERS, H. S. The Morphology of *Aspergillus herbariorum*. Ann. Myc. v, p. 419.
1908 DOMARADSKY, M. Zur Fruchtkörperentwickelung von *Aspergillus Fischeri* Wehm. Ber. d. deut. bot. Ges. xxvi A, p. 14.
1909 DALE, E. On the Morphology of *Aspergillus repens* de Bary. Ann. Myc. vii, p. 215.
1909 SCHIKORRA, W. Ueber die Entwickelungsgeschichte von *Monascus*. Zeitschr. für Bot. i, p. 379.
1910 THOM, C. Cultural Studies of Species of *Penicillium*. U.S. Dept. Agr. Bureau Animal Industry Bull. 118.

### *Onygenaceae*

The Onygenaceae include only the remarkable genus *Onygena*. There are some six species, all limited in habitat to such animal substances as horns, hoofs, feathers, fur and skin. This peculiarity, together with the absence of conidia, the thin wall of the perithecium, and the fact of its dehiscence by lobes or by a circular split, distinguishes the members of the group from the other Plectascales, which they resemble in the irregular arrangement of their asci. Two of the six species have sculptured spores and sessile fructifications which thus approach those of the Aspergillaceae, while in the other four the spores are smooth and the fructification stalked. To the latter group belongs *Onygena equina*, described by Marshall Ward in 1899. The ascospores germinate only after a prolonged resting period or after treatment with gastric juice; thus treated they produce a vigorous mycelium on which the ascocarp first appears as a dome-shaped mass of white hyphae; a little later it becomes covered with loose cells among which air is entangled, rendering the fruit very difficult to wet. As development proceeds a number of hyphae grow outwards and divide into short segments, certain of which swell up and are liberated as chlamydospores, covering the whole of the stroma with a dense powder.

Meantime the hyphae which gave rise to these become crowded together to form the pseudoparenchymatous sheath. Internally asci are produced and give rise each to eight spores, but the ascus walls soon disappear and the spores lie free amongst the vegetative hyphae. This mature stage, in which there is no trace of asci, caused *Onygena* to be variously classified with the Myxomycetes and with the Lycoperdaceae before its true position was discovered. There is no evidence of the existence of sexual organs.

## ONYGENACEAE: BIBLIOGRAPHY

1899 WARD, H. MARSHALL. *Onygena equina* Willd. A Horn-destroying Fungus. Phil. Trans. B. cxci, p. 269.

1917 BRIERLEY, W. B. Spore Germination in *Onygena equina*, Willd. Ann. Bot. xxxi, p. 17.

### *Elaphomycetaceae and Terfeziaceae*

In the next two families, Elaphomycetaceae and Terfeziaceae, the fruit is subterranean. The species differ from the other hypogeal Ascomycetes, the Tuberales, with which they are still sometimes classified, and resemble the subaerial Plectascales in the irregular arrangement of their asci, which are scattered or grouped in nests surrounded by sterile branches (fig. 37). The gleba or central complex of hyphae is not at any stage of development in communication with the exterior.

In the Elaphomycetaceae the ascocarp is surrounded by a thick yellow or brown peridium, the asci are subglobose and the gleba breaks up at maturity into a powdery mass of spores. The only genus is *Elaphomyces*. The mycelium of certain species develops in relation to the roots of *Pinus* and other conifers, and the ascocarp is often parasitized by species of the pyrenomycetous fungus *Cordyceps*. *E. granulatus*, the commonest British species, is the host of *C. capitata*, and *E. variegatus* of *C. ophioglossoides* (fig. 110).

In the Terfeziaceae (figs. 36, 37) the peridium is much less distinct, and

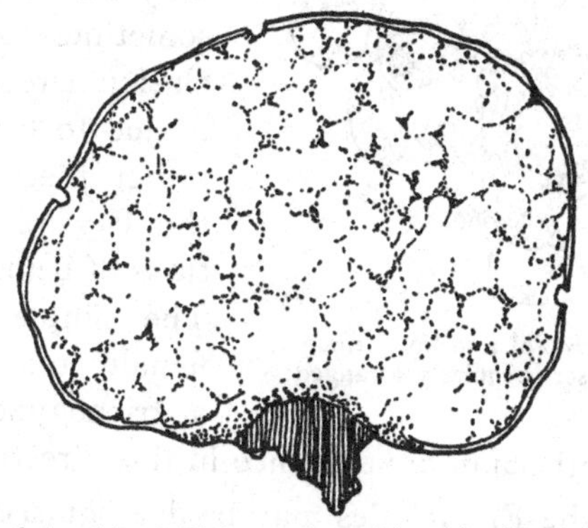

Fig. 36. *Terfezia olbiensis* Tul.; section of fructification; after Tulasne.

in some cases is represented merely by an ascus-free region around the periphery of the fruit. Moreover the spores do not, as in the Elaphomycetaceae, form a powdery mass at maturity.

## ERYSIPHALES

The Erysiphales are characterized by an abundant superficial mycelium, which may be white (colourless) or dark-coloured. The perithecia are spherical, ovoid or flattened, and are usually without an ostiole; the peridium is thin and membranous; the asci are arranged in a regular layer at the base of the perithecium.

The group includes some 600 species, the majority of which are external parasites or epiphytes upon the leaves of higher plants. They are grouped into three families, of which the Microthyriaceae are but little known, and of doubtful position, and the Erysiphaceae and Perisporiaceae show several points in common both with the Plectascales, from which they differ in the regular arrangement of their asci, and with the Pyrenomycetes, from which they are for the most part distinguished by the absence of an ostiole.

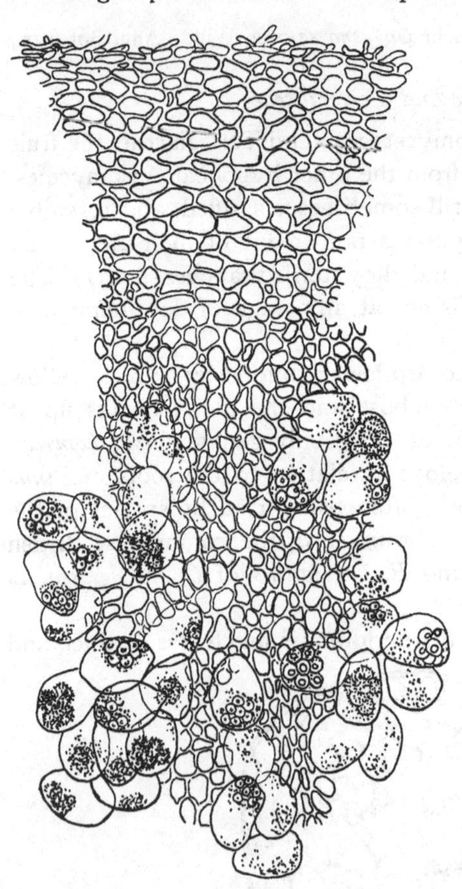

Fig. 37. *Terfezia olbiensis* Tul.; section through hymenium, showing asci irregularly arranged; after Tulasne.

Their taxonomic position is probably somewhere between these two groups, and they have, under various systems of classification, been placed in closer proximity sometimes to the one and sometimes to the other. Their inclusion here in the Plectomycetes is due to the fact that they, or rather their best-known family, the Erysiphaceae, show indications of being a primitive group. The simple type of male and female organs, the latter without a trichogyne, and the simple structure of the perithecium are evidence in this direction.

The families of the Erysiphales may be distinguished as follows:

Aerial mycelium colourless (or white). Perithecia more or less globose without an ostiole, furnished with conspicuous appendages.
Conidia of oidium type. ERYSIPHACEAE.

Aerial mycelium dark-coloured or rarely absent. Perithecia globose or ovoid, without appendages.
Conidia not of oidium type. PERISPORIACEAE.
Aerial mycelium dark-coloured or absent. Perithecia flattened or shield-shaped, with an ostiole at the apex, without appendages.
Conidia absent. MICROTHYRIACEAE.

A further and probably important distinction which separates the Erysiphaceae from the other two families is the character of the ascus and ascospores. In *Erysiphe* and its allies the ascus is more or less globose, the spores are always continuous and colourless, and the number of spores in the ascus is frequently reduced. In the Perisporiaceae, on the other hand, the ascus is relatively elongated and sometimes cylindrical, and the spores are commonly two or more celled and often dark-coloured. Most species agree, however, with the Erysiphaceae in the absence of paraphyses. The Microthyriaceae approach the Perisporiaceae rather than the Erysiphaceae in the characters of the ascus and spore, but, as already stated, they are clearly distinguished by the curious flattened perithecium.

It remains for future research to determine whether these families should be grouped together, or whether the Perisporiaceae and Microthyriaceae are true Pyrenomycetes and should be placed in that alliance. In our present extensive ignorance of the initiation and development of their perithecia, it seems unwise to remove them from their traditional position in the neighbourhood of the Erysiphaceae.

## *Erysiphaceae*

The members of the Erysiphaceae are popularly known as white or powdery mildew or blight. They have a practically worldwide distribution, but have been recorded especially in Europe and the United States.

They are obligate parasites on the leaves or young shoots and inflorescences of flowering plants. The germinating conidium or ascospore gives rise to an abundant, superficial, septate mycelium of uninucleate cells, which ramifies over the leaf in every direction, forming a white web-like coating, and sends haustoria into the epidermal cells of the host. In the simplest cases the haustorium is a slender tube which swells up inside the host cell; in other species it is branched, sometimes forming finger-like processes, and frequently there is an external disc or swelling (adpressorium), from which, or from the mycelium in the neighbourhood of which, the haustorium proper arises and pushes into the epidermal cell. As a rule the fungus does not penetrate further, but in *Erysiphe Graminis* Salmon was able to induce endophytic growth and nutrition by allowing the conidia to germinate on wounded leaves; in *Phyllactinia Corylea* the branches of the aerial mycelium enter the stomata, extend through the intercellular spaces and send haustoria into the neighbouring cells, and in *Erysiphe* (or *Oidiopsis*) *taurica* the whole mycelium during the conidial stage is located in the tissues of the host. We have thus, within the limits of the family, a transition between ecto- and endoparasitism through hemiendophytic forms, and forms which are endophytic under abnormal conditions. When perithecia are about to be produced and the mycelium emerges and spreads over the surface of the host leaf, the hyphae both of *Phyllactinia* and of *E. taurica* show haustorial branches (adpressoria), though no haustoria are produced. It may be inferred that the ectophytic condition with haustoria penetrating the epidermal cells is primitive in the group.

Indian and Persian specimens of *E. taurica* have been found under practically desert conditions, others have been collected on plants of the steppes of Turkestan at a height of 6000 feet, and in localities exposed to very dry winds. The suggestion has consequently been made that the endophytic habit in this family is an adaptation to xerophytic conditions, since it both provides shelter for the developing mycelium and obviates the necessity of piercing through the cuticle, which in desert plants is of considerable thickness.

The Erysiphaceae are propagated during the summer by rather large oval uninucleate conidia (fig. 38). These are ordinarily produced in rows on simple conidiophores with one or more basal cells. In the endophytic *E. taurica*, however, the conidia are borne singly on branched conidiophores which emerge through the stomata of the host.

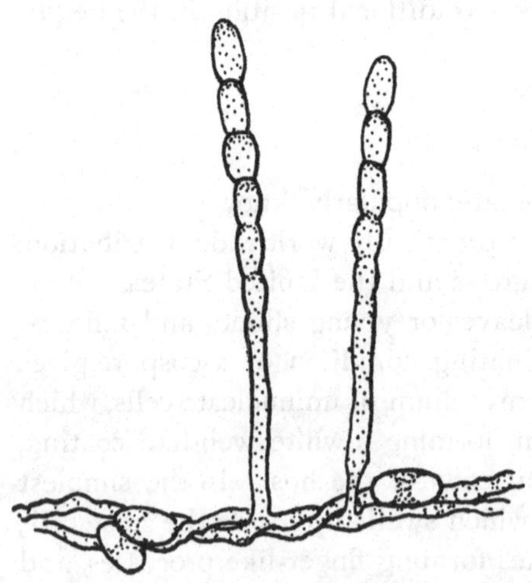

Fig. 38. *Sphaerotheca pannosa* Wallr.; conidiophores and conidia, × 240.

In the case of *Phyllactinia Corylea*, which is met with on a large number of deciduous trees, variations occur in the shape of the conidia borne on different hosts, and indicate the existence of morphological distinctions between the biological forms of the species.

Before the connection between the conidia and the perithecia of the Erysiphaceae was understood, the generic name, *Oidium*, was applied to the former. The name is still used to indicate the characteristic form of the conidial stage and to describe conidia when the perithecia are unknown. This was the case with the powdery vine mildew. The conidial form, known as *Oidium Tuckeri*, became prevalent in Europe in 1845–6, but perithecia were not observed till 47 years later, when they appeared during two successive seasons in various localities in France, and the fungus was identified with the American vine mildew, *Uncinula necator*, in which perithecia are common. The unusual production of the perithecial stage was attributed to the sudden alternation of high and low temperatures which characterized the seasons in question. The survival of the fungus in the absence of ascospores has been attributed to the persistence of the mycelium, and also to the development of conidia capable of passing the winter in the resting state. Fortunately the disease is readily kept in check by the application of appropriate sprays.

A very similar case is that of the oak mildew. In or about 1904, oak scrub in England and many parts of Europe became infected with the conidial form *O. Quercinum*; in 1911 perithecia were found in France, and the parasite was identified with the common American form *Microsphaera Alni* which frequently occurs on oaks in the United States. Here again exceptional seasonal conditions appear to have been necessary for the formation of the perithecial stage.

In a different manner climate has affected the development of the gooseberry mildew, *Sphaerotheca mors-uvae*, which was introduced into Europe from America about 1900. Very numerous perithecia are developed but a considerable proportion either fail to mature or fail to survive the winter. Infection of the young shoots in the spring appears to depend on the earliest formed perithecia, which alone have had time to mature and lodge in the bark or between the bud scales of the host. For this reason prevention is here much more difficult than in the case of the vine mildew, since the mature perithecia are very difficult to kill and the spread of the disease must be combated by the removal of infected shoots in the autumn, or by appropriate methods of cultivation.

The perithecia of the Erysiphaceae appear in the late summer or autumn; they are spherical or subglobose, 50—300 $\mu$ (0·05 to 0·3 mm.) in diameter and furnished with simple or branched appendages; they are fixed in position by means of a secondary mycelium. When quite young the perithecia appear white and glistening like the vegetative mycelium; later they become clear yellow and finally brown in colour. Their development is by no means simultaneous, so that a considerable range of stages can often be seen within the field of an ordinary hand lens.

During development the wall of the perithecium is differentiated into inner and outer layers (fig. 39). The inner layer is several cells thick; its cells are rich in cytoplasm with thin, apparently unmodified walls; it is in contact with the developing asci, about which it forms a packing, and to which it supplies food material. Outside this layer is a strengthening and protective zone of several series of cells with scanty contents, the walls of which undergo a change apparently analogous to lignification. In *Phyllactinia* the outermost layer, from which the secondary mycelium and the characteristic appendages are derived, consists of thin-walled cells, but in other genera it is not differentiated from the protective zone.

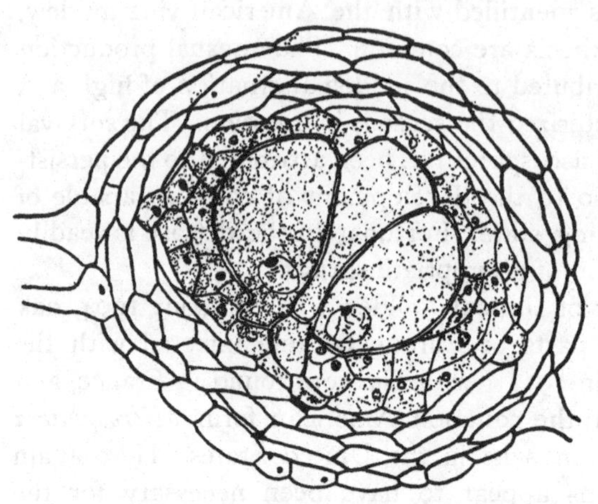

Fig. 39. *Erysiphe Polygoni*; young perithecium containing uninucleate asci; after Harper.

A single ascus or several may be formed in the perithecium; the ascospores, numbering two to eight in each ascus, begin to develop during the summer or autumn, but they remain in the perithecium under normal conditions and do not germinate till the beginning of the following year, when they are set free by the rupture of the perithecium wall, and produce the first infections of the season. In *Erysiphe Graminis* and *Sphaerotheca mors-uvae* some of the perithecia separate readily from the mycelium in August, and, if supplied with moisture, will eject their spores after a few hours. They may in this way prove a source of infection in the season in which they were produced, and this is probably true of other species also, and should be borne in mind if ripe ascospores are being searched for.

The family includes six genera, all of which are British, and are readily distinguished (fig. 40).

In *Sphaerotheca* (one ascus), and in *Erysiphe* (several asci), the perithecial appendages are filamentous, unbranched or branched irregularly, and very like the ordinary hyphae of the mycelium; in *Podcsphaera* (one ascus), and in *Microsphaera* (several asci), they are usually dichotomously branched; in *Uncinula* the apices of the appendages are spirally coiled, and in *Phyllactinia* the perithecium bears stiff, pointed hairs with swollen bases. In this genus also the apex of the perithecium is furnished with a ring of richly branched penicillate cells. At about the time of spore-formation these break down, forming a sticky gelatinous cap, by means of which the perithecium, after it is first set free, adheres upside down to its original host plant or to other objects. In view of this peculiarity the ascription of *Phyllactinia* to any host in contact with which its perithecia may be found demands careful verification.

The function of the true appendages in the Erysiphaceae is somewhat variable. In *Phyllactinia* the bulb at the base of the appendage is thick-walled over its upper surface, but an oval region remains thin on the lower side. As the ripening perithecium loses water so do the appendages; the thin area below the bulb is pushed in by atmospheric pressure, the under surface becomes consequently shorter than the upper and the end of the spine is pulled down till subsequent moistening straightens it again (Harper).

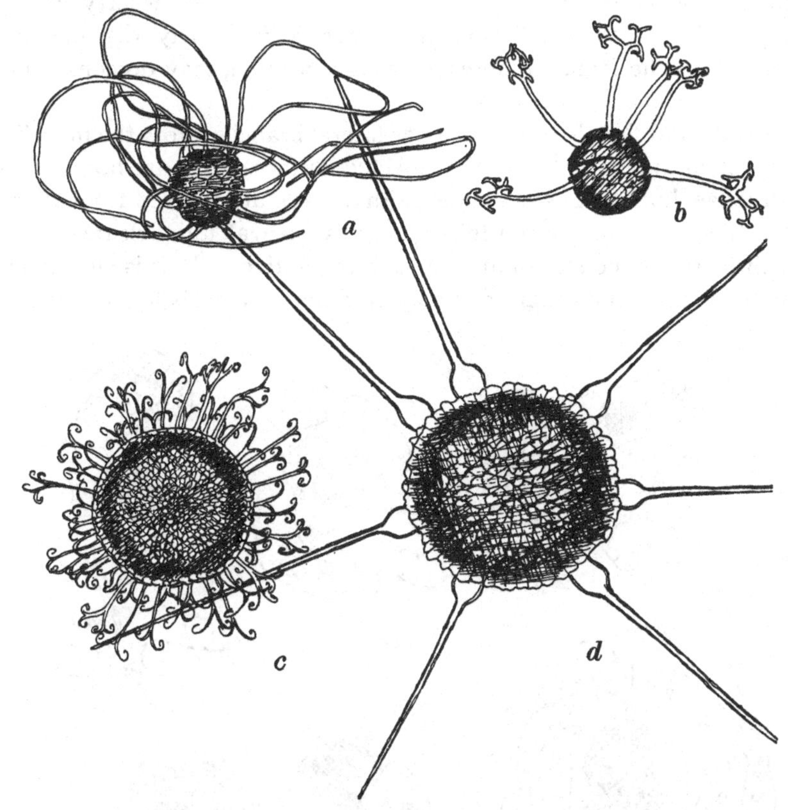


Fig. 40. Perithecia of *a. Erysiphe tortilis* (Wallr.) Fr.; *b. Microsphaeria sp.*; *c. Uncinula Aceris* (DC.) Sacc.; *d. Phyllactinia Corylea* (Pers.) Karst.; × 120.

These hygroscopic movements may be repeated many times according to weather conditions, even after the living protoplast has disappeared from the appendage (Neger); and at last the perithecium is loosened from its attachment.

In other cases, such as *Erysiphe* and *Sphaerotheca*, the appendages may help to anchor the perithecium to its host during development; in *Uncinula necator* their apices become mucilaginous (Salmon), and they serve, much as do the penicillate cells of *Phyllactinia*, to attach the free perithecium upside down to the substratum.

The development of the perithecium was first described in 1863 by de Bary, who was able to recognize an antheridium and oogonium and the formation of an ascus or asci from the latter. These and several subsequent investigations have rendered the reproductive processes in the Erysiphaceae better known than perhaps in any other group of fungi.

*Sphaerotheca Humuli*[1] occurs on a variety of common plants, on the cultivated strawberry, where it is responsible for strawberry mildew, and especially on the hop. On the latter it is widely distributed in autumn, and, if the female inflorescences are infected, may do considerable damage.

The male and female organs arise as lateral branches from the mycelium, and project at right angles to the infected surface; they are borne on different hyphae, but there is no evidence that these are derived from distinct mycelia. The oogonium, when fully grown, is an oval, uninucleate structure two or three times the size of an ordinary vegetative cell; it is cut off from the parent hyphae and a stalk cell may be differentiated below it (fig. 41 *a*).

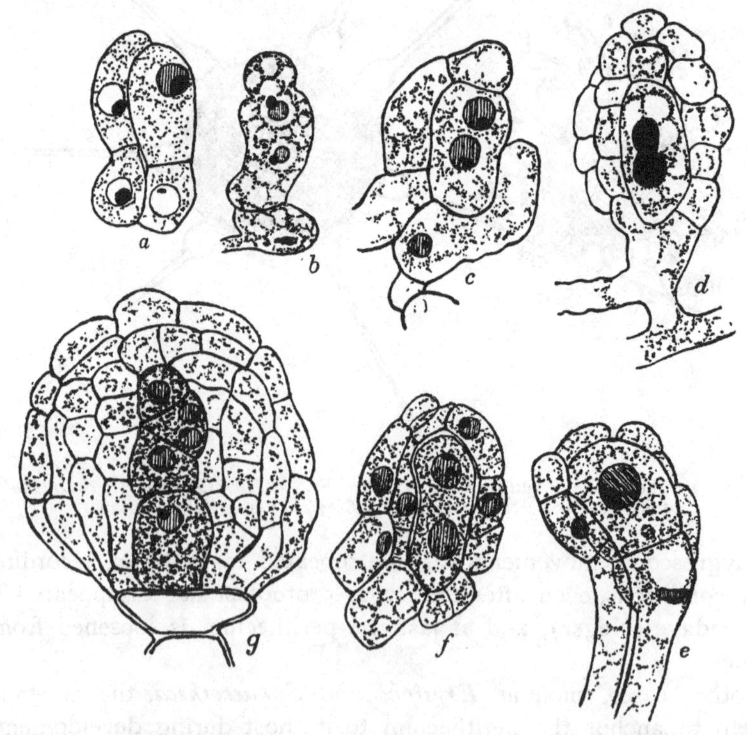


Fig. 41. *Sphaerotheca Humuli* (DC.) Burr.; *a*. young oogonium and antheridium; *b*. entrance of male nucleus; *c*. male and female nuclei in oogonium; *d*. fertilization; *e*. fusion nucleus; *f*. nuclei produced by first division of fusion nucleus; *g*. young perithecium with binucleate ascogenous cell; × 1360; after Blackman and Fraser.

[1] *S. Humuli* (DC.) Burr. = *S. Castagnei* Lev.

The antheridial branch is much narrower; it applies itself to the side of the oogonium and when first cut off contains a single nucleus (fig. 41 *a*). It is clearly differentiated from the hyphae of the sheath not only by its form and behaviour but by its much earlier appearance and definite relation to the oogonium. Its nucleus soon divides; one of the daughter nuclei passes to the apex of the branch and a wall is formed cutting off the uninucleate antheridium. The oogonial nucleus is rather larger than those of the vegetative cells, the antheridial nucleus decidedly smaller.

The investigations of Harper, and subsequently of Blackman and Fraser, show that the wall between the oogonium and antheridium now breaks down (fig. 41 *b*), the male nucleus passes into the oogonium, travels to the female nucleus and fuses with it (fig. 41 *c*, *d*). The pore between the sexual cells soon closes and the cytoplasm left behind in the antheridium degenerates, sometimes forming a densely staining cap on the oogonium. The fertilized oogonium elongates, the fusion nucleus divides (fig. 41 *f*) and a wall separates the daughter nuclei so that two uninucleate cells are formed. The upper divides again and eventually a row of cells is produced, the penultimate containing two nuclei while the others are uninucleate (fig. 41 *g*). The formation of the first wall is sometimes delayed, so that the undivided oogonium may for a time show two or three nuclei.

At about the time when the sexual organs reach maturity the development of the sheath begins, branches grow out from the stalk of the oogonium and enclose the sexual organs in a single layer of cells. These give off branches towards the interior, and thus a second zone is formed whose cells become rich in cytoplasm and contain two or three nuclei each. The nutritive and protective envelopes are thus laid down and their further development produces the typical sheath already described.

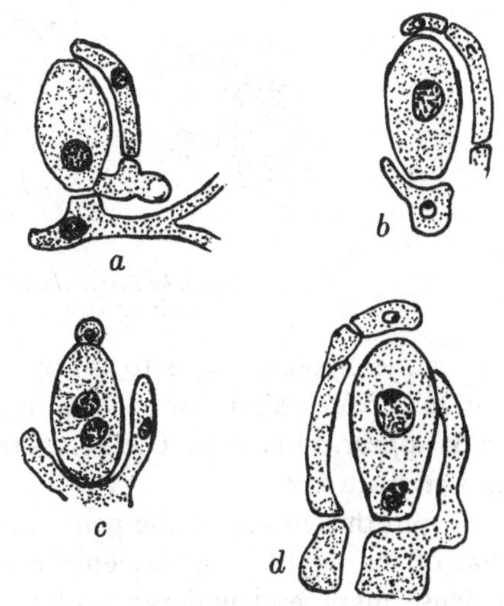


Fig. 42. *Sphaerotheca Humuli* (DC.) Burr.; development of archicarp; in *c*. two nuclei, regarded as the product of division, are shown in the oogonium, while a cell at the top of the oogonium, regarded as the antheridium, still contains a nucleus; after Winge.

Meantime the penultimate cell of the septate oogonium enlarges to form the single ascus characteristic of *Sphaerotheca*, the terminal cell and the cells below the ascus are pushed aside and disappear, the two nuclei fuse, the usual three successive nuclear divisions take place, and ascospores are produced.

The only record of the number of

chromosomes is that of Winge who describes eight in the first and second mitoses, and four in the third, and suggests that a brachymeiotic reduction takes place.

According to the interpretations of Dangeard and Winge fertilization does not take place in the oogonium of *Sphaerotheca Humuli*. For them the degenerating mass in the antheridium includes the male nucleus which thus degenerates *in situ* and the two nuclei seen lying side by side in the female organ represent the product of a premature division (fig. 42).

Again Bezssonoff working on *Sphaerotheca mors-uvae*, the economically important gooseberry mildew, records the entrance of the male nucleus into the oogonium, but does not observe its fusion with the female nucleus. He finds four chromosomes throughout the divisions in the ascus.

*Erysiphe Polygoni*[1] occurs on the leaves and stems of a considerable variety of hosts belonging to a number of different families. The development of the sexual organs takes place much as in *Sphaerotheca Humuli*, and, here also, Harper has observed the entrance of the male nucleus into the oogonium, and its fusion with the female nucleus (fig. 43*a*). After fertilization the

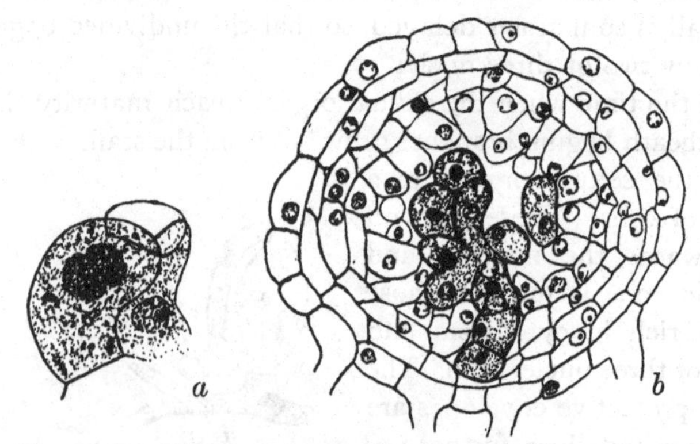


Fig. 43. *Erysiphe Polygoni*; *a*. fertilization; *b*. young perithecium with ascogenous hyphae; after Harper.

protective hyphae begin to grow up, the oogonium elongates, the fusion nucleus divides till a row of from five to eight nuclei is produced, transverse walls appear, and a row of cells is formed of which the penultimate contains two or more nuclei.

From the surface of the penultimate cell, and perhaps sometimes from that of its neighbours, filaments bud out (fig. 43*b*), branch rapidly to form a dense mass, and undergo septation. These are the ascogenous hyphae. In them some six or eight intercalary cells, which will give rise to asci, become distinguished by the fact that each contains two nuclei. The rest

[1] *Erysiphe Polygoni* DC. = *Erysiphe communis* (Wallr.) Link and Rabh., and *E. Martii* Lév.

lose their contents and are displaced by the developing asci. Later the fusion of the two nuclei in each ascus takes place, and in each eight spores are formed.

Dangeard, investigating the development of *E. Polygoni* and *E. Cichoracearum*, notes that in his material the oogonium underwent septation before a row of nuclei was formed, and that cells other than the penultimate contained two or more nuclei. Usually in *E. Cichoracearum* and sometimes in *E. Polygoni* the oogonial branch consisted of two cells; this corresponds with the arrangement in the antheridial branch, which is regularly bicellular. In both cases the lower cell is to be regarded as a stalk. In regard to the occurrence of fertilization Dangeard's conclusions correspond with those which he reached in relation to *Sphaerotheca.*

*Phyllactinia Corylea* infects the leaves of deciduous trees and shrubs including ash, oak, beech, hazel and hornbeam.

The sexual organs arise (Harper 1905), as in other mildews, where two hyphae intersect. They become closely applied to each other, and, as the oogonium grows more quickly than the antheridial branch, it becomes somewhat twisted around the latter. The subsequent history is very like that of *Sphaerotheca* or *Erysiphe*. A uninucleate antheridium is cut off, the male nucleus enters the female organ (fig. 44*a*), nuclear fusion takes place, the

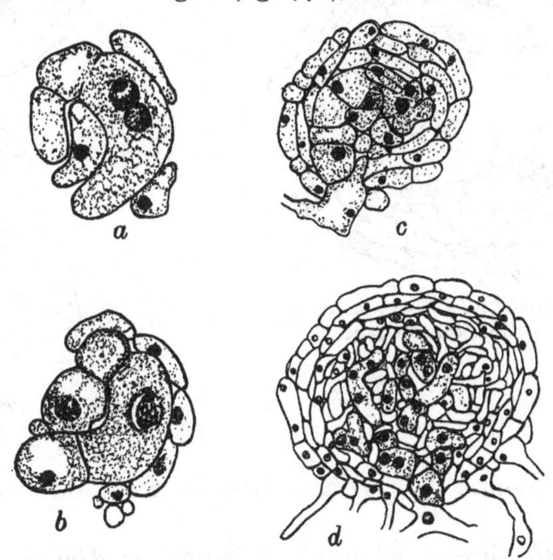


Fig. 44. *Phyllactinia Corylea* (Pers.) Karst.; *a*. fertilization; *b*. fusion nucleus in oogonium; *c*. *d*. young perithecia; after Harper.

oogonium elongates and enlarges in diameter and the fusion nucleus divides. The first nuclear division is apparently never accompanied by cell wall formation, so that a binucleate stage persists for some time. Finally, however, the usual row of three to five cells is formed. The penultimate cell regularly contains more than one nucleus; the rest, as a rule, are uninucleate.

Just after fertilization the sheath begins to grow up (fig. 44 *b*), springing in this case from the stalk cell of the antheridium, as well as from that of the oogonium, and developing into the three layers described above.

The ascogenous hyphae arise as lateral branches from the septate oogonium (fig. 44 *c*), all or most being derived from the penultimate cell about which they are crowded and intertwined. They are at first multinucleate, and, as development proceeds, push up vertically within the perithecium (fig. 45); septation then takes place. The asci, of which there are several in each perithecium, arise as lateral outgrowths from the intercalary cells, or are formed directly from the terminal cells of the ascogenous hyphae. Each young ascus contains two nuclei, but the remaining cells are almost without exception uninucleate. Fusion takes place in the ascus (fig. 46) and is followed by three nuclear divisions; as a rule only two spores are formed.

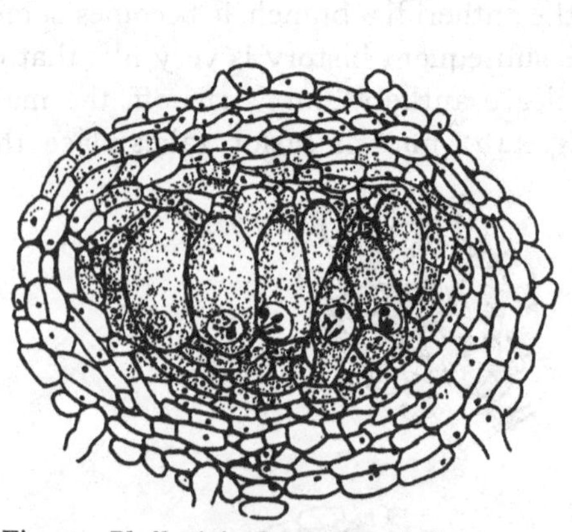

Fig. 45. *Phyllactinia Corylea* (Pers.) Karst.; perithecium containing uninucleate asci; after Harper.

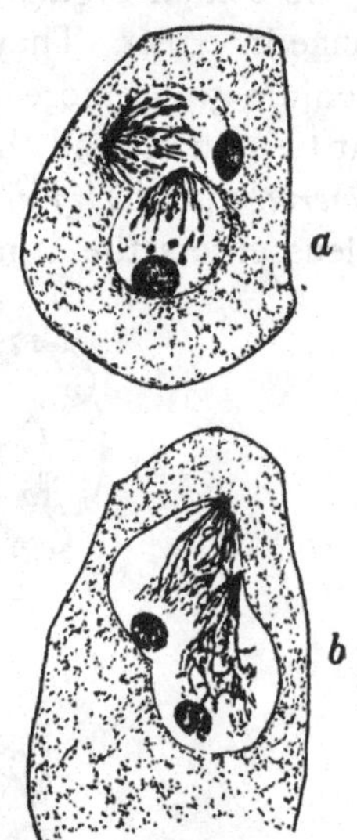


Fig. 46. *Phyllactinia Corylea* (Pers.) Karst.; *a*. *b*. fusion in ascus; after Harper.

Eight chromosomes (fig. 47) have been observed throughout the life-history.

In *Phyllactinia Corylea* and also in *Microsphaera Alni* (Sands, 1907) and various species of *Erysiphe* (Harper, 1905), the organization of the resting nucleus is very characteristic. A deeply staining **central body** lies against the nuclear membrane and to this the chromatin threads are attached. From it they extend into the central cavity of the nucleus forming a sheaf of divergent rays connected laterally by delicate fibrillae (fig. 46).

In *Phyllactinia* there is evidence that the number of main strands corresponds to the number of chromosomes, and that, in fact, these persist

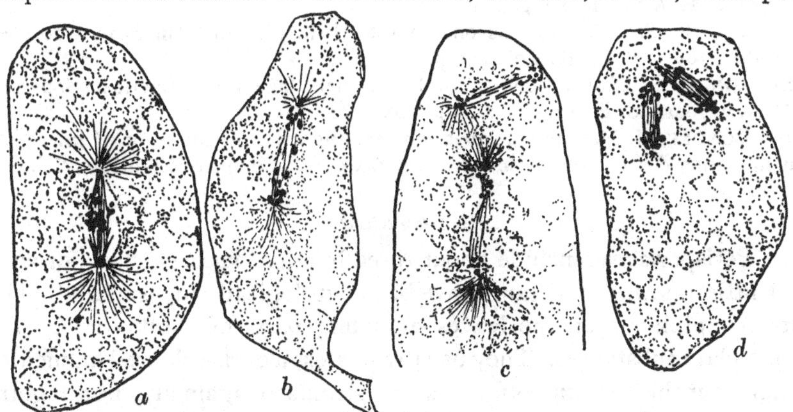


Fig. 47. *Phyllactinia Corylea* (Pers.) Karst.; *a.* metaphase of first division in ascus; *b.* anaphase of first division; *c.* anaphase of second division in ascus; *d.* anaphase of third division; after Harper.

throughout the resting stages that intervene between successive divisions, and fuse in pairs after nuclear fusion.

## ERYSIPHACEAE: BIBLIOGRAPHY

1895 HARPER, R. A. Die Entwickelung des Peritheciums bei *Sphaerotheca Castagnei.* Ber. d. deutsch. Bot. Gesell. xiii, p. 475.

1896 HARPER, R. A. Über das Verhalten der Kerne bei der Fruchtentwickelung einiger Ascomyceten. Jahrb. für wiss. Bot. xxix, p. 655.

1897 DANGEARD, P. A. Second mémoire sur la production sexuelle des Ascomycètes. Le Botaniste, v, p. 245.

1897 HARPER, R. A. Kernteilung und freie Zellbildung im Ascus. Jahrb. für wiss. Bot. xxx, p. 249.

1900 SALMON, E. S. A Monograph of the Erysiphaceae. Mem. Torrey Bot. Club, ix, p. 1.

1903 NEGER, F. W. Neue Beobachtungen über das spontane Freiwerden der Erysipheen-fruchtkörper. Centr. f. Bakt. Parasit. u. Infekt. ii, Abt. x, p. 570.

1905 BLACKMAN, V. H. and FRASER, H. C. I. Fertilization in *Sphaerotheca.* Ann. Bot. xix, p. 567.

1905 HARPER, R. A. Sexual Reproduction and the Organisation of the Nucleus in Certain Mildews. Publ. Carn. Inst. Washington.

1905 SALMON, E. S. On Endophytic Adaptation shown by *Erysiphe graminis* D.C. under cultural conditions. Phil. Trans. cxcviii, p. 87.

1906 SALMON, E. S. On *Oidiopsis taurica* an Endophytic member of the Erysiphaceae. Ann. Bot. xx, p. 187.

1907 DANGEARD, P. A. L'origine du périthèce chez les Ascomycètes. Le Botaniste, x, p. 216.

1907 SALMON, E. S. Notes on the Hop Mildew. Journ. Agr. Sci. ii, p. 327.

1907 SANDS, M. C. Nuclear Structure and Spore Formation in *Microsphaera Alni.* Trans. Wisconsin Acad. Sci., Arts and Letters xv, pt. 2, p. 733.

1911 WINGE, O. Encore le *Sphaerotheca Castagnei.* Bull. Soc. Myc. de France, xxvii, p. 211.

1912 FOËX, M. Les conidiophores des Erysiphacées. Rev. Gén. Bot. xxiv, p. 200.
1913 SALMON, E. S. Spraying Experiments against the American Gooseberry Mildew, Journ. S. E. Ag. Col. xxii, p. 404.
1913 SALMON, E. S. Observations on the Life-History of the American Gooseberry Mildew, Journ. S. E. Ag. Col. xxii, p. 433.
1913 SALMON, E. S. Observations on the Perithecial Stage of the American Gooseberry Mildew, Journ. S. E. Ag. Col. xxii, p. 440.
1914 BEZSSONOFF, M. N. Sur quelques faits relatifs à la formation du périthèce et la délimitation des ascospores chez les *Erysiphaceae*. Comptes Rendus Ac. Sci. clviii, p. 1123.

*Perisporiaceae*

The Perisporiaceae include about three hundred species, many of which are but little known, while none have been cytologically investigated. They develop as epiphytes on the leaves or young parts of plants, or occur on decaying plant substances. They usually possess a dark-coloured filamentous mycelium, but this sometimes forms a firm stroma, or again may be altogether lacking. The perithecia are superficial, dark in colour, and usually more or less spherical; they are typically without appendages, but mycelial outgrowths from their base may simulate these structures as in *Meliola*. The wall of the perithecium is generally membranous, more rarely carbonaceous and brittle; as a rule there is no definite opening but sometimes an irregular rent is formed at the apex (*Antennaria*), and sometimes the perithecium opens by valves (*Capnodium*). The asci are elongated and more or less cylindrical and the spores have one or more septa and are sometimes muriform; paraphyses are not generally developed.

*Dimerosporium*, the largest genus with some sixty species, is epiphytic on the leaves of angiosperms, and one species (*D. Collinsii*) forms witches'-brooms on the service-berry. The mycelium is dark-brown, the spores two-celled.

The species of *Capnodium*, *Apiosporium* and *Meliola*, are among the soot fungi, which form a black coating on leaves. They are purely epiphytic and saprophytic, subsisting on the honey dew secreted by insects, and doing little damage, as they are seldom thick enough to interfere with the supply of light.

In several species a considerable variety of accessory fructifications are produced. Thus *Meliola Penzigi*, the "sooty-mould" of the orange, has conidia which differ little from the vegetative cells, multi-cellular conidia, conidia borne in small spherical pycnidia, and conidia abstricted from conidiophores in pustules or conceptacles, which may be flask-shaped or variously branched; some of these accessory spores are developed in great abundance and perithecia are relatively rare.

PERISPORIACEAE: BIBLIOGRAPHY

1897 WEBBER, H. J. Sooty Mould of the Orange and its Treatment. U.S. Dept. Ag. Veg. Phys. and Path. Bull. 13.
1892 ELLIS, J. B. and EVERHART, B. M. North American Pyrenomycetes. Ellis and Everhart, New Jersey.

### *Microthyriaceae*

The aerial mycelium of the Microthyriaceae is dark-coloured and superficial; the flattened perithecia are shield-shaped and only the upper part of the sheath is fully developed; it consists of a disc of radiating hyphae, which increase by branching as they grow towards the periphery, and are firmly attached one to another along their lateral walls. In this way a continuous layer of pseudoparenchyma is formed, below which the asci develop more or less at right angles to the surface of the leaf. The asci are cylindrical or pyriform, and the spores frequently bicellular. A definite ostiole may be formed, as in *Microthyrium*, or the perithecium may tear open at the apex as in *Asterina*. These two, with about forty and ninety species respectively, and *Asterella* with sixty are the largest genera in the family. The species are mainly tropical with a few representatives in Europe and North America.

### MICROTHYRIACEAE: BIBLIOGRAPHY

1913 THIESSEN, F. Über einige Mikrothyriaceen. Ann. Myc. xi, p. 493.
1914 THIESSEN, F. Über Membranstructuren bei die Microthyriaceen. Myc. Centr. iii, p. 273.

## EXOASCALES

The Exoascales form a group of obligate parasites on vascular plants; they cause hypertrophy of the infected parts, producing yellow, red or purple discolorations, blisters, curling of the leaves, malformation of the fruit, and sometimes abnormal branching with the formation of tufts of fasciated twigs known as witches'-brooms. The latter peculiarity is by no means always attributable to the Exoascales but is induced also by certain rusts and by the attacks of insect parasites. The Exoascales are responsible for several diseases of economic importance including peach leaf curl induced by *Exoascus deformans*, a witches' broom on cherries due to *E. Cerasi*, and the distortion known as pocket plums caused by the presence of *E. Pruni*, which infects the flesh of the fruit and inhibits the development of the stone.

Infection apparently takes place at about the time of the opening of the buds of the host, probably by means of spores deposited on the bud-scales; in cold, moist weather, when the young leaves are in a state of lowered vitality, the fungus readily gains entrance; it can be checked by the use of appropriate sprays. Once in the leaf the hyphae in most cases ramify between the cells of the host, but in *Taphrinopsis Laurencia* on *Pteris biaurita* they are intracellular. No haustoria are developed.

The mycelium may be annual or it may be perennial, hibernating mainly in the cortex and medulla of the young twigs and causing hypertrophy, especially of the hypodermal tissue, so that infected branches appear abnormally thick. In such cases the destruction of the infected parts is necessary in order to combat the disease.

The fertile mycelium is richly branched and consists of relatively short cells; it is developed mainly on the leaves or carpels of the host, and in

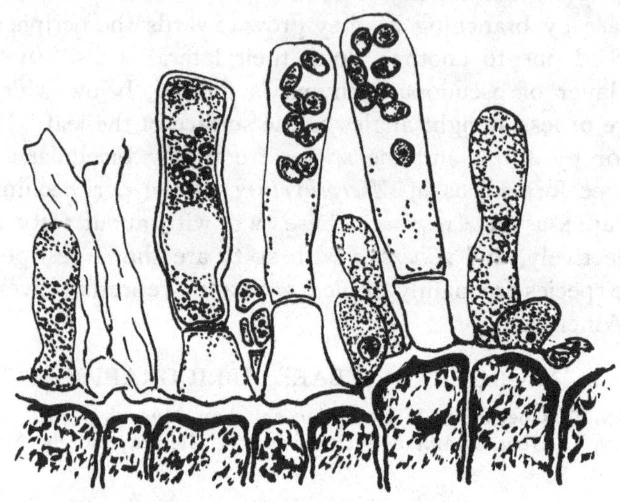

Fig. 48. *Exoascus deformans* (Berk.) Fuck., × 1000.

these regions asci are produced. Sometimes the mycelium permeates the whole tissue and the asci arise from hyphae below the epidermal cells and push up among them (*Taphrina aurea*), sometimes the fertile hyphae lie between the epidermal cells (*Magnusiella Potentillae*), but in the majority of cases the asci are developed above the epidermal cells and just below the cuticle (*Exoascus deformans*). In the species of *Taphrinopsis* occurring on *Pteris* the asci are produced within the epidermal cells.

The asci either arise directly from the mycelium (figs. 49, 50) or each is borne on a small, cubical stalk cell (fig. 48) which is cut off from the ascus

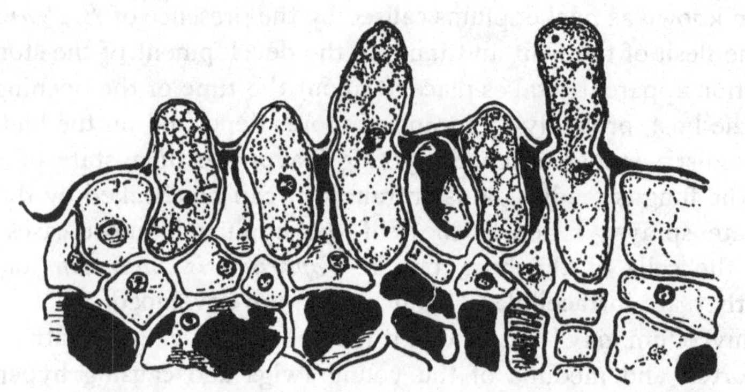

Fig. 49. *Taphrina aurea* (Pers.) Fr.; young asci, × 500.

mother-cell during development. Two nuclei can frequently be recognized in the cells of the fertile mycelium, and the young ascus, in all investigated cases, is binucleate. The two nuclei fuse, the fusion nucleus undergoes three successive divisions and eight spores are formed (fig. 48). In many species budding of the ascospore takes place, so that the mature ascus contains numerous minute conidia (fig. 50) by means of which the fungus is distributed.

The Exoascales include the single family Exoascaceae; with this is sometimes associated the Ascocorticiaceae containing the saprophytic

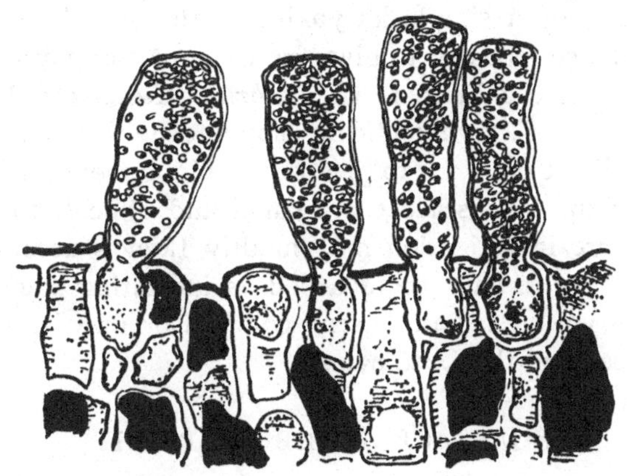

Fig. 50. *Taphrina aurea* (Pers.) Fr.; mature asci, × 500.

genus *Ascocorticium* with five known species. The asci of *Ascocorticium*, like those of the Exoascaceae, are cylindrical in form, parallel in arrangement and quite unprotected. It is open to question whether the parallel arrangement of the asci in the Exoascaceae has any phylogenetic significance or is not rather the result of their development on the surface of the carpel or leaf. The cylindrical form of the ascus, however, does not suggest a primitive group. It may perhaps be inferred that the Exoascaceae are reduced forms derived from one of the phyla with protected asci; there does not appear at present to be any clue to their probable ancestry. The same is true of the Ascocorticiaceae, of the development of which even less is known.

### *Exoascaceae*

The classification of the Exoascaceae, and especially the separation of the two principal genera, *Exoascus* and *Taphrina*, has been based on a variety of different characters including the form of the ascus, the development or not of conidia from the ascospores, the annual or perennial character of the mycelium and the presence or absence of a stalk cell. Boudier defines *Exoascus* as having asci usually octosporous and usually provided with a basal cell; *Taphrina* as having asci usually polysporous, sometimes provided with a basal cell.

The mycelium on which the asci are borne consists wholly or mainly of binucleate cells. The young ascus, as shown by Dangeard in 1894, is at first binucleate; the nuclei soon fuse and the fusion nucleus divides in preparation for spore-formation.

In *Taphrina Cerasi* Ikeno in 1903 described the presence, after nuclear fusion, of a densely staining nucleolus or chromatin body as the only nuclear structure in the ascus. A spindle was formed, apparently from the substance of the nucleolus, the remainder of which became the single chromosome. The latter divided by a simple karyokinesis, the spindle was reabsorbed and the process twice repeated to give rise to eight chromatin bodies about which the spores were delimited. In *Taphrina Kusanoi* and other species Ikeno found no sign of karyokinetic division but the chromatin body underwent successive direct divisions to give rise to eight spore nuclei.

The nuclei of the Exoascaceae are small and difficult to stain so it is possible that future investigation may modify Ikeno's account; should it be confirmed it may perhaps be regarded as indicating stages in the disappearance of karyokinesis and a useful comparison may be instituted with the similar processes in the Uredinales.

## EXOASCACEAE: BIBLIOGRAPHY

1887 KNOWLES, E. L. The Curl of Peach Leaves; a Study of the Abnormal Structure produced by *Exoascus deformans*. Bot. Gaz. xii, p. 216.

1894 DANGEARD, P. A. La reproduction sexuelle des Ascomycètes. Le Botaniste, iv, p. 30.

1903 IKENO, S. Die Sporenbildung von *Taphrina*-Arten. Flora, xcii, p. 1.

# CHAPTER IV

## DISCOMYCETES

THE term Discomycetes is applied to those groups in which the fruit is more or less cup-shaped (fig. 51), with the hymenium fully exposed at maturity, and to their immediate allies. The ascocarp is surrounded by a peridium or sheath of closely interwoven hyphae which is closed at first and later is pushed apart by the paraphyses, so that at last it forms the outer

Fig. 51. *Otidea aurantia* Mass.; apotheci, nat. size.

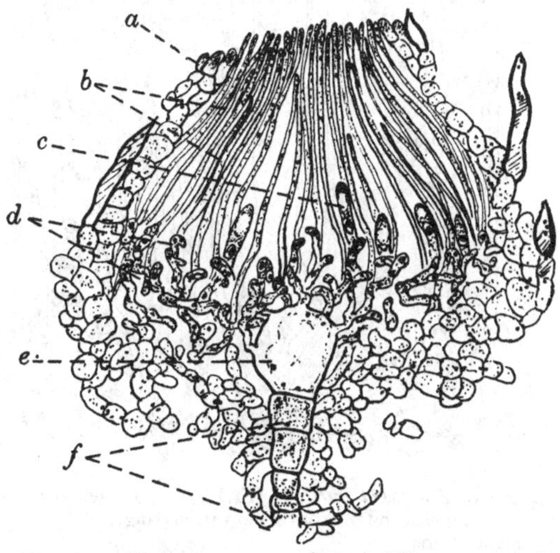


Fig. 52. *Lachnea stercorea* (Pers.) Gill.; ascocarp in longitudinal section showing young asci and paraphyses, × 160. *a*. sheath; *b*. paraphyses; *c*. ascus; *d*. ascogenous hyphae; *e*. oogonium; *f*. stalk of archicarp.

wall of the cup (fig. 52). The lower part of the cup is filled by the **hypothecium**, a tangle of hyphae, some vegetative, some ascogenous. These give rise to the **sub-hymenial layer** where the paraphyses have their origin and where the young asci are developed. The asci and paraphyses grow up together and rise to the surface of the ascocarp forming the **hymenium** or fertile disc which is spread over the interior of the cup. The asci are more or less cylindrical and parallel one to another and to the paraphyses (fig. 53). They open either by a lid (fig. 55) or by the ejection of a plug (fig. 54). They arise in succession so that large numbers may be produced in a single ascocarp. If the hypothecium is well developed the apothecial cup is full and the hymenium lies across the brim like the skin on a bowl of custard; if the development of the hypothecium is slight the hymenium spreads over the sides and bottom of the cup. In many cases, as in *Peziza vesiculosa* and *Otidea aurantia*, the cup is small and comparatively full when it first opens, and grows larger and deeper as development proceeds.

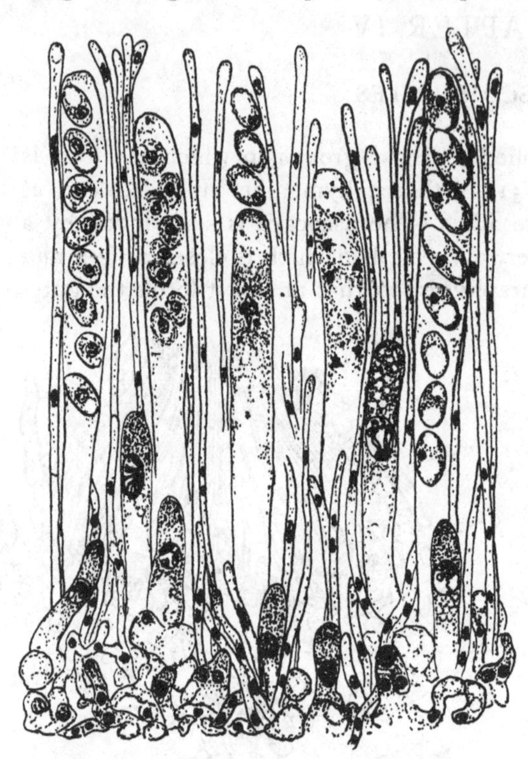

Fig. 53. *Humaria rutilans* (Fr.) Sacc.; hymenial layer showing asci and paraphyses in various stages of development, × 400.

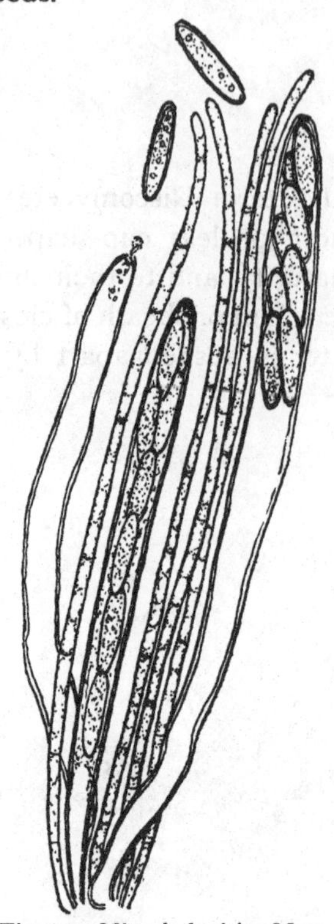

Fig. 54. *Mitrula laricina* Mass.; development and ejection of biseriate spores, × 600.

This typically discomycetous ascocarp or apothecium, which is well seen in the Pezizales, may be connected in one direction, through the Patellariaceae and their allies, with the fructifications of the Phacidiales, which are partly closed with a more or less stellate aperture, and with the characteristically elongated fructifications of the Hysteriales, which open by a narrow slit. The apothecia of this last series show a close resemblance to the perithecia of certain Pyrenomycetes, and, as far as their mature structure is concerned, they may be placed as logically in one group as the other. Nor have we at present any information with regard to development which can decide the question. It remains even possible that the Hysteriales are a transition series and that some of the forms grouped among Discomycetes may have arisen from the Pyrenomycetes or *vice versa*.

In another direction the typical apothecium, when very widely open, suggests the reflexed ascocarp of *Rhizina* or *Sphaerosoma* and such a type may, by invagination of the fertile surface, have produced the closed fruit of the truffles. The simpler Tuberales may have had a similar origin, or may have arisen direct from a pezizaceous form, such as *Sepultaria*, with which *Genea* has several points in common.

It is not impossible that the *Rhizina* group, by the development of a sterile stalk, has also produced the Helvellaceae and it may be the Geoglossaceae as well. But the latter family, because of the characteristic method of dehiscence of its ascus (by the ejection of a plug of wall material), has sometimes been looked upon as related to the Helotiaceae and their allies which show a similar dehiscence.

Massee has developed a theory that forms with large, coloured, and elaborately sculptured spores, tend to be primitive. He thus regards the Tuberaceae and Geoglossaceae as ancient groups from both of which pezizaceous forms have sprung.

Bucholtz's work on the development of *Tuber* diminishes the probability that this is a primitive type, or one that has given rise to cup-shaped forms, and it seems easier to think of *Genea* and its allies as derived from the Pezizales by the diminution in size of the external aperture, the shortening and broadening of the ascus and the increased convolution of the hymenium, than to regard them as giving rise to that group by the contrary changes. It is, however, not improbable that the Pezizales are polyphyletic in origin, and that some of them may have been derived from the higher Geoglossaceae.

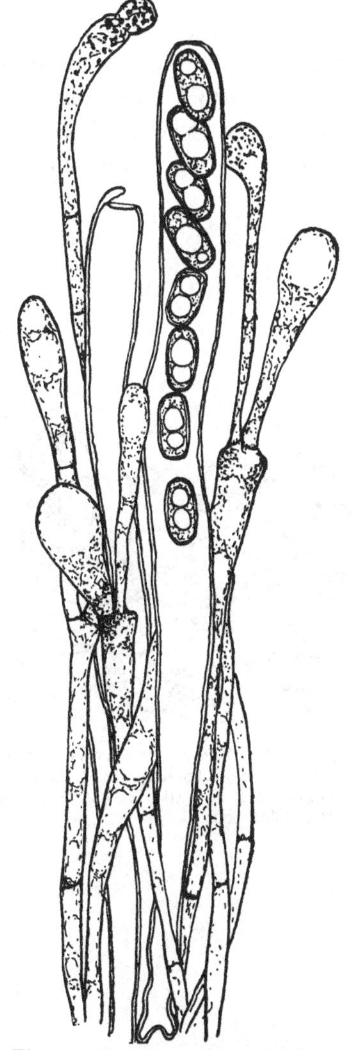

Fig. 55. *Sepultaria coronaria* Mass.; uniseriate spores; ascus opening by a lid; branched, septate, clavate paraphyses; ×600.

So far we have considered mainly the external characters of the fruit and the structure of the ascus, but when we turn to the Pezizales we find a further, and possibly a more valuable, criterion in the structure of the sexual organs.

The antheridium is known in very few cases. A large, oblong coenocytic cell has been described in *Ascodesmis*, and a similar, though larger, organ in *Pyronema* and in *Lachnea stercorea*, and we have also, if its position should be ultimately established, the curious stalked conidium of *Ascobolus carbonarius*.

The archicarp is of much commoner occurrence, and seems more likely to be useful as a gauge of relationship. Among Discomycetes the simplest type is undoubtedly that of *Ascodesmis* or *Thelebolus*; the significant details in *Thelebolus* are not fully known, but in *Ascodesmis* we have a stout, twisted hypha, divided into three parts, the unicellular trichogyne, the unicellular coenocytic oogonium and the multicellular stalk (fig. 56). After fertilization

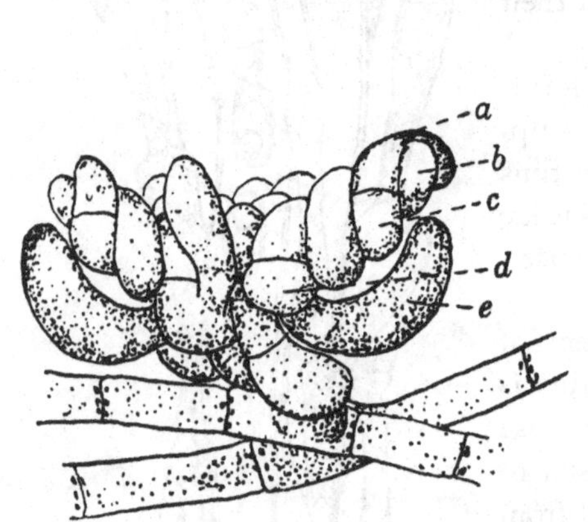


Fig. 56. *Ascodesmis nigricans* Van Tiegh.; sexual apparatus; *a.* trichogyne; *b.* antheridium; *c.* oogonium; *d.* stalk; *e.* gametophytic hypha; after Claussen.

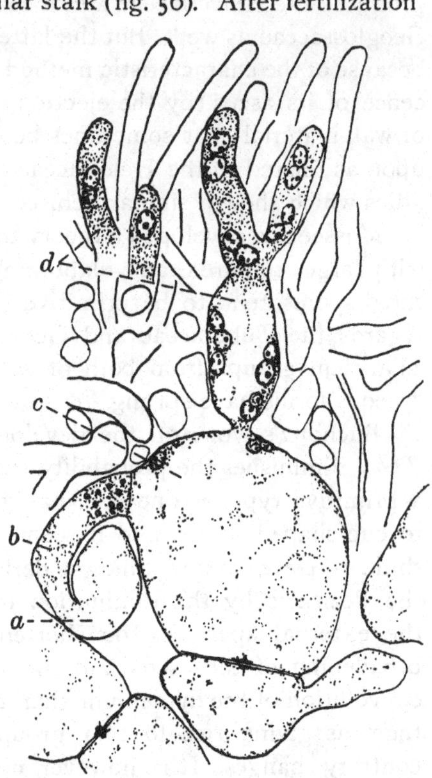


Fig. 57. *Pyronema confluens*; spherical oogonium giving rise to ascogenous hyphae; *a.* antheridium; *b.* trichogyne; *c.* oogonium; *d.* ascogenous hyphae; × 1040; after Claussen.

the oogonium becomes septate, so that the fertile part is multicellular and the ascogenous hyphae arise from several cells. This type closely approximates to that in *Eurotium* and some other Plectascales, and there seems reason to regard it as a primitive female organ among Discomycetes.

From it may be derived the spherical oogonium of *Pyronema* (fig. 57) which differs mainly in the fact, no doubt connected with its shape, that it does not become septate after fertilization, so that the ascogenous hyphae arise from one cell only. The same is true of *Lachnea stercorea*, which differs in its multicellular trichogyne, and of *Humaria granulata*, and perhaps some other forms in which the trichogyne, like the antheridium, has disappeared.

Another group is distinguished by the multicellular oogonial region of the archicarp, which has also a stalk and a terminal region (or trichogyne) of several cells each (fig. 58). This type of female organ is sometimes termed a **scolecite**. It occurs in *Lachnea cretea* and in several species of *Ascobolus* and *Ascophanus*.

Fig. 58. *Ascobolus furfuraceus* Pers.; archicarp, × 740; after Dodge.

In *Ascobolus furfuraceus* several of the central cells communicate one with another by means of pores, but only one of them gives rise to ascogenous hyphae; in some other species of *Ascobolus* and in the genus *Ascophanus* ascogenous hyphae arise from several communicating cells; the same is true of *Lachnea cretea*, where three or four cells become practically continuous owing to the disappearance of the transverse septa between them.

In the trichogyne also, or terminal region of the archicarp, pores are formed in the transverse walls, so that this multicellular organ offers no bar to the passage of male nuclei.

Such structures may well have been derived from the *Ascodesmis* type by the elongation and transverse septation of the parts of the archicarp. Our knowledge of the sexual organs of the Pezizales thus suggests that they may have been derived, along two principal lines of development, from a common ancestor within the group.

In only two other families of Discomycetes, the Rhizinaceae (*Rhizina* and *Sphaerosoma*) and the Geoglossaceae (*Leotia*) has an oogonium been recognized. In each of these a large fertile cell is present, but its development is not known and no suggestions as to phylogeny can therefore be based upon it.

Our knowledge of the development and especially of the reproductive organs of the Discomycetes is still very incomplete and further research is very necessary. As has been shown in *Leotia*, *Humaria* and other cases, it by no means follows that because one member of a genus has lost its sexual apparatus the same will be true of others. Every available species should be investigated.

The Discomycetes include over 4000 species and may be subdivided as follows:

Hymenium fully exposed at maturity
 mature ascophore cup-shaped — PEZIZALES.
 mature ascophore reflexed or stalked with the fertile region often convoluted — HELVELLALES.

| | |
|---|---|
| Hymenium incompletely exposed at maturity | |
| ascophore round, aperture usually stellate | PHACIDIALES. |
| ascophore elongated, opening by a slit | HYSTERIALES. |
| Hymenium enclosed at maturity | TUBERALES. |

## PEZIZALES

The Pezizales are characterized by the fleshy or sometimes leathery ascocarp, bounded, except in the Pyronemaceae, by a more or less definite peridium which is closed at first and opens later, giving the mature fruit a cup or saucer shape.

This dehiscence is due to the growth of a conical mass of paraphyses which push out at the apex of the ascocarp. Fresh paraphyses and, a little later, asci grow up amongst those first formed and the peridium is pushed wide open and the hymenium exposed.

The asci contain usually eight, but sometimes four, sixteen, thirty-two, or numerous spores, which germinate typically by means of germ-tubes, but which, in a few cases. give rise by budding to conidia. Various accessory spores, including conidia, chlamydospores and oidia, are also produced.

The mycelium is well developed and filamentous, rarely forming a sclerotium. The species are parasitic or saprophytic on the ground or on dead parts of plants; in many cases they are coprophilous.

Considering the very large size of the group the number of species investigated is small. In a few of these, normally functional male and female organs have been found, in some the antheridium has disappeared, and in many the oogonium is also lacking. Where an oogonium is present it gives rise to the ascogenous hyphae, while the paraphyses originate from the stalk or from the surrounding cells (fig. 52).

The group may be divided into the following families:

| | |
|---|---|
| Peridium continuous with hypothecium | |
| Peridium incomplete; ascocarps usually compound | PYRONEMACEAE. |
| Peridium well-developed | |
| asci not rising above the surface when ripe; ascospores usually uniseriate | PEZIZACEAE. |
| asci rising above the surface when ripe; ascospores often coloured and biseriate | ASCOBOLACEAE. |
| Peridium distinct from hypothecium | |
| Peridium of elongated hyphae (pseudoprosenchymatous) | HELOTIACEAE. |
| Peridium pseudoparenchymatous | MOLLISIACEAE. |
| Peridium absent or ill-defined; epithecium formed | CELIDIACEAE. |
| Peridium tough; epithecium formed | |
| ascocarp free | PATELLARIACEAE. |
| ascocarp embedded when young | CENANGIACEAE. |
| Apothecia numerous, sunk in a stroma | CYTTARIACEAE. |

*Pyronemaceae*

The Pyronemaceae are a small group distinguished from the other Pezizales by the fact that the peridium, or lateral boundary of protective hyphae around the fruit, is not well developed.

This is not always regarded as a sufficiently important character to warrant their separation from the Pezizaceae and many authors include them in that group.

The only important genera are *Ascodesmis* and *Pyronema*, species of both of which have been somewhat fully investigated.

*Ascodesmis nigricans*[1] (fig. 59) is a small coprophilous form.

The sexual organs appear in artificial culture about forty-eight hours after the germination of the spore. Stout, multinucleate hyphae grow up from the mycelium and dichotomize (fig. 60*a*) to give rise to some six or eight archicarps. Near these, and usually from the same filament, one or two antheridial hyphae arise. They grow towards the archicarps (fig. 60*b*) and dichotomize (fig. 60*c*), while around each of their terminal cells or antheridia, an archicarp becomes wrapped (fig. 60*d*).

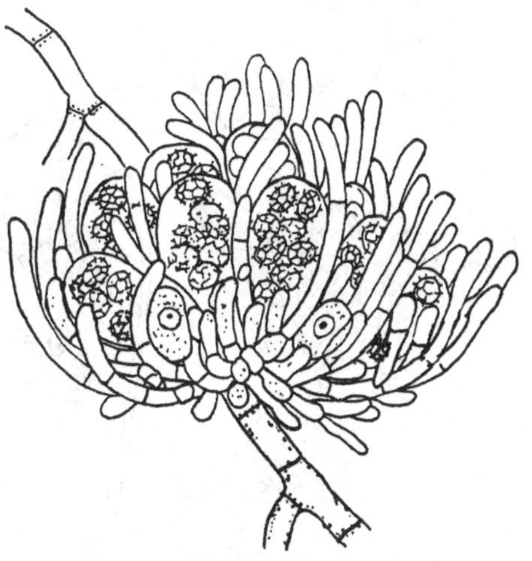

Fig. 59. *Ascodesmis nigricans* Van Tiegh.; apothecium, × 340; after Claussen.

In the meantime walls are laid down, so that the various archicarps and antheridia become cut off from their neighbours, and each archicarp divides transversely to form a trichogyne and an oogonium. The trichogyne usually contains two nuclei, the oogonium five or six and the antheridium about the same number. The nuclei of the trichogyne soon degenerate, and, as observed by Claussen, the wall between this cell and the antheridium is broken down (fig. 60*e*), so that open communication is established. The male nuclei pass into the oogonium, where for a time ten or twelve nuclei may be counted, then fusion of these in pairs takes place (fig. 60*f*). Subsequently the oogonium enlarges somewhat and undergoes septation; large ascogenous hyphae, usually about three in number, bud out from it (fig. 60*g*), and quickly give rise to asci (fig. 60*h*). Ascus formation is apparently quite typical, the spores are spherical and have a characteristically sculptured epispore (fig. 60*i*). At

[1] Claussen described the cytology of this species under the name of *Boudiera hypoborea* Karst.; see Cavara, *Ann. Myc.* iii, 1905, p. 363, and Dangeard, *Botaniste*, x, 1907, p. 247, for nomenclature.

about the time of fertilization vegetative filaments begin to grow up (fig. 60*d*), and at last form a loose investment around and among the developing asci (fig. 59).

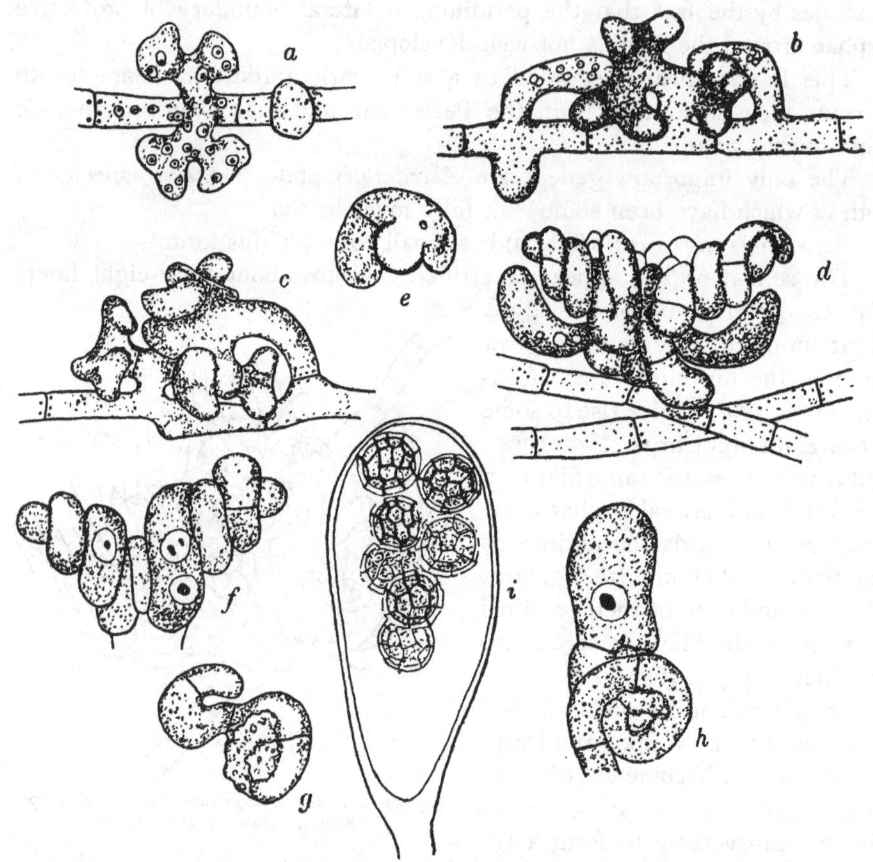


Fig. 60. *Ascodesmis nigricans* Van Tiegh.; *a. b. c. d.* development of the sexual apparatus; *a.* and *b.* × 1000, *c.* × 1100, *d.* × 800; *e.* communication between antheridium and trichogyne, × 1300; *f*. fusion in oogonium, × 1600; *g.* septate oogonium and ascogenous hypha; antheridium and trichogyne shrivelled, × 1000; *h.* uninucleate ascus, × 1100; *i.* sculptured spores in ascus, × 750; after Claussen.

*Pyronema confluens*[1] occurs on burnt ground or on charred and decayed leaves in woods. Its fruits are pink or salmon-coloured, its mycelium to a great extent superficial, and its sexual apparatus of unusually large size. The latter fact led to an early study of its development.

It was described by de Bary in 1863, by the brothers Tulasne in 1865 and 1866, by van Tieghem in 1884, and by Kihlman in 1885, and a very clear understanding of the morphology of the sexual organs was reached. A swollen, elongated antheridium was recognized and a more or less globular

[1] *Pyronema confluens* Tul. = *P. omphaloides* (Bull.) Fuckel.

oogonium, from which a trichogyne protruded (fig. 61 *b*). The union of the trichogyne and antheridium was observed and it was shown that from the oogonium ascogenous hyphae subsequently arose. Van Tieghem recorded that the species is very susceptible to external conditions, the antheridium sometimes being reduced in size or absent, though the oogonium nevertheless developed normally and produced ascogenous hyphae.

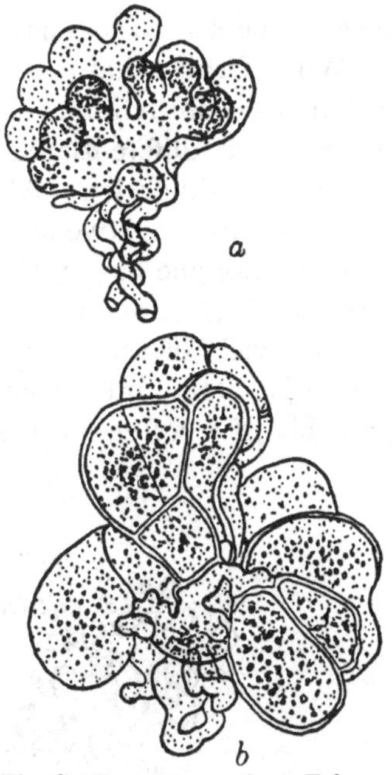

Fig. 61. *Pyronema confluens* Tul.; *a*. development of sexual apparatus; *b*. mature oogonia and antheridia; × 390; after de Bary.

In 1900 appeared the classical researches of Harper, followed in 1903 and 1907 by Dangeard's, and in 1912 by Claussen's investigations.

The mycelium is made up of cells of varying length, regularly multinucleate, containing six to twelve nuclei, and filled with cytoplasm of a loose, spongy structure. The mycelium is sparse and loose, but the reproductive organs are very abundant so that the ascocarps, when mature, are often crowded together.

The first indication of sexual reproduction consists in the appearance of thick hyphae, very similar to the corresponding filaments in *Ascodesmis*, and tending, like them, to stand at right angles to the substratum. They dichotomize (fig. 61 *a*), the terminal cells become swollen, and two types, the more spherical oogonium and the more elongated antheridium, are distinguished; these arise on separate branches, but from the same mycelium. Both are multinucleate from their initiation.

Very soon a slight elevation appears on the oogonium; it elongates rapidly to form the multinucleate trichogyne and, before its growth is complete, is separated from the oogonium by a wall; the trichogyne and antheridium grow towards one another, the tip of the trichogyne meeting sometimes the apex, but more commonly the flank of the male organ.

The mature oogonium is a spherical or flask-shaped cell filled with dense cytoplasm and containing many nuclei which are very much larger than those of the ordinary vegetative cells. Its stalk consists of two or three broad cells, its apex is continued into the trichogyne. The nuclei of the latter increase but little in size and are thus much smaller than those of the essential organs at maturity. The nuclei of the antheridium are almost as large as those of the female cell, but its protoplasm is less dense owing perhaps to the absence of accumulated reserve materials. It would appear that in both oogonium and antheridium some nuclei degenerate (Claussen). Hyphae from the ascogonial and antheridial branches, and also from the surrounding cells, begin to grow up even before fertilization, and later envelop the sexual organs.

When fertilization is about to take place an area of cytoplasm in the region of the antheridium, against the wall of which the tip of the trichogyne is pressed, is differentiated as a very finely granular disc from which the nuclei are withdrawn. Although located in the antheridium this area resembles the receptive spot seen in the oosphere of many algae. The tip of the trichogyne has by this time developed as a beak-like projection, and this also is empty of nuclei and contains dense and finely granular cytoplasm.

The walls of the antheridium and trichogyne now break down at the point of contact and a pore is formed. The process is gradual, consisting probably of a softening and solution of the wall material, which seems to

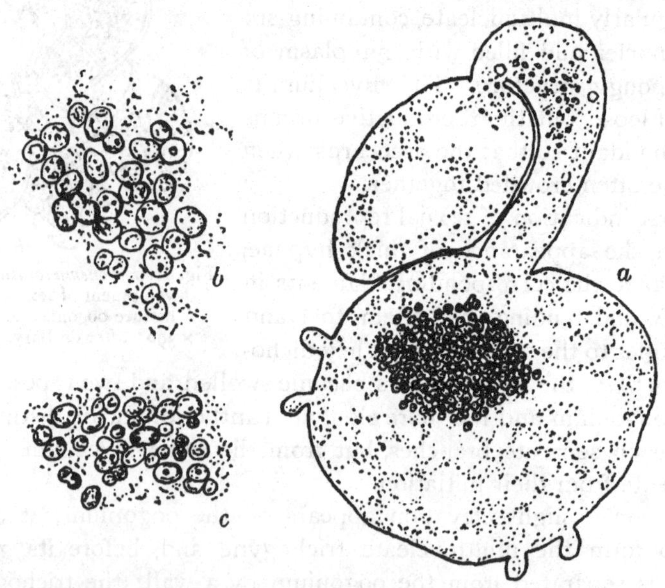


Fig. 62. *Pyronema confluens*; *a.* antheridium, trichogyne and oogonium, male and female nuclei collected in the middle of the latter; *b. c.* fusion of male and female nuclei; after Harper.

spread out into the cytoplasm of the beak, suggesting that the solvent action is mainly exerted from the interior of the trichogyne. The open pore now becomes thickened around its margin so that an exceedingly strong ring unites the antheridium and trichogyne, and they can be bent or turned upon each other without being pulled apart. This arrangement is no doubt necessary to withstand the strain set up by the flow of nuclei from the relatively wide cavity of the antheridium through the narrow pore and beak.

While the formation of the pore is in progress the nuclei of the trichogyne degenerate, and, by the time that they are completely disorganized, a migration of the male nuclei through the pore begins. Ultimately the contents of the trichogyne degenerate still further, till the cytoplasm and nuclei together form a densely staining mass which may be recognized even in the mature fruit. The male nuclei continue to pass into the tube until it is densely filled, and sometimes a trifle swollen. According to most observers the wall at the base of the trichogyne now breaks down so that an open passage is formed and the male nuclei travel through (fig. 63*a*), and mingle

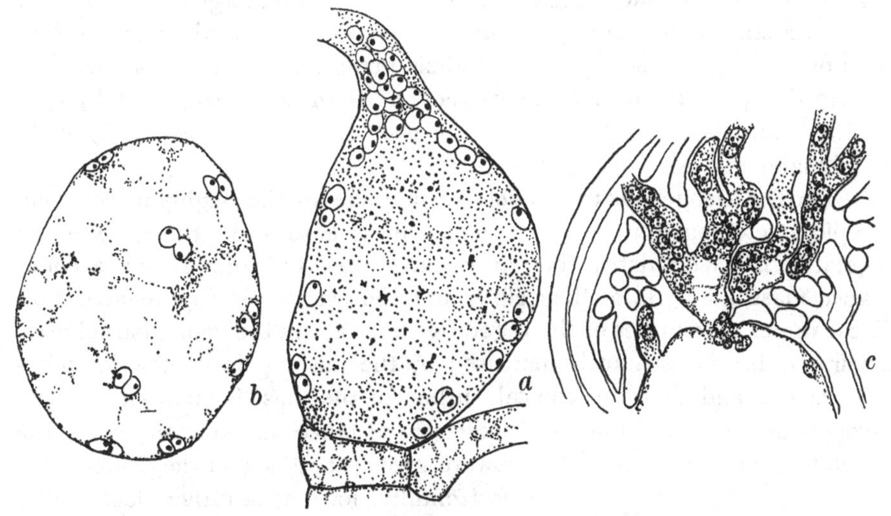


Fig. 63. *Pyronema confluens*; *a*. entrance of male nuclei into oogonium, × 1435; *b*. association of male and female nuclei, × 1160; *c*. ascogenous hyphae with nuclei in pairs, × 820; after Claussen.

with the female nuclei. After their migration is complete a fresh wall is laid down across the base of the trichogyne, cutting off the oogonium once more as a single spherical cell.

The female nuclei meantime become aggregated together (fig. 62*a*) and form a hollow sphere or dome or a sickle-shaped group. This is no doubt a provision for insuring their association with the male nuclei with the greatest certainty and dispatch.

According to both Harper and Claussen, the sexual nuclei now pair; Harper has recorded complete fusion at this stage (fig. 62*b*), while Claussen (fig. 63*b*) regards the nuclei as merely associated in preparation for their ultimate union in the ascus. Dangeard, on the other hand, denies the disappearance of the wall between the oogonium and trichogyne or the passage of the male nuclei beyond the latter organ, and Brown has described a variety in which the trichogyne and antheridium fail to unite.

In any case the greater part of the male cytoplasm does not enter the oogonium but is left behind in the antheridium and trichogyne; consequently these organs, after their function is complete, remain, to the superficial view, unchanged for a long period, till they are crushed at last by the growth of the investing hyphae, or perhaps destroyed by bacteria (Harper, p. 354).

Before the stages described above, the oogonium has begun to bud out (fig. 62*a*) at various points, giving rise to the ascogenous hyphae. Into these the nuclei pass, a few, no doubt unpaired, being left behind in the oogonium. The hyphae elongate, branch freely and undergo septation, and, as the vegetative filaments grow up, they ramify among them and at last bend over and give rise to asci from their penultimate cells. Claussen has described a paired arrangement of the nuclei in the ascogenous hyphae (fig. 63*c*), and believes the members of each pair to be respectively male and female.

After the ascogenous and vegetative hyphae are thoroughly interwoven, a rapid stretching upward of the whole mass ensues. In this growth the vegetative hyphae outstrip the reproductive ones, and form at first a cone-shaped mass, made up of their elongated, slender, densely aggregated tips. These upper extremities of the vegetative hyphae are the young paraphyses. Their number is constantly increased by the pushing in of new branches from below, and thus the conical outline of the mass is maintained. The ascogenous hyphae grow for a certain distance in company with the vegetative filaments, then their upward growth ceases, and they spread out horizontally, forming a rather dense layer below the cone of paraphyses. This is the base of the hymenium.

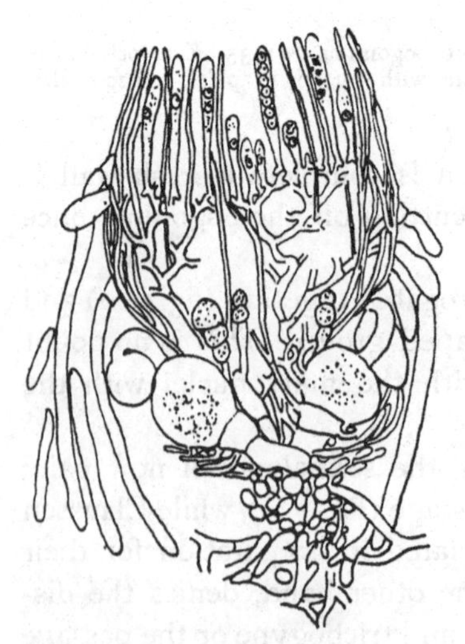

Fig. 64. *Pyronema confluens*; diagrammatic section through ascocarp; after Harper.

Usually in *Pyronema*, as in *Ascodesmis*, several oogonia are invested by a common sheath, and their ascogenous hyphae mingle to form the hymenium of a single ascocarp (fig. 64), but ascocarps developed in relation to a single pair of sexual organs are not unknown.

The formation of the asci in *Pyronema* is quite typical. The number of chromosomes is probably ten (Harper), or twelve (Claussen), at any rate in the first divisions in the ascus. Dangeard records a smaller number in the third division, and in the variety *inigneum* Brown describes five throughout.

As development proceeds the sexual organs become completely crushed and are at last no longer recognizable. At an early stage it becomes impossible to trace the connection between the ascogenous hyphae and the oogonium, and, during a great part of their development, these depend for their nutrition upon the paraphyses and other vegetative cells.

A secondary mycelium grows downwards to the substratum, obtaining food material from it and serving for the attachment of the mature ascocarp. Special storage cells appear in the hypothecium.

## PYRONEMACEAE: BIBLIOGRAPHY

1863 DE BARY, A. Entwickelungsgeschichte der Ascomyceten. Leipzig.

1865 TULASNE, L. R. and C. Selecta fungorum Carpologia, iii. Imperial. typograph., Paris.

1866 TULASNE, L. R. and C. Note sur les phénomènes de copulation que présentent quelques Champignons. Ann. Sci. Nat. vi, p. 217.

1884 VAN TIEGHEM, Ph. Culture et développement du *Pyronema confluens*. Bull. Soc. Bot. de France, xxxi, p. 355.

1885 KIHLMAN, O. Zur Entwickelungsgeschichte der Ascomyceten. *Pyronema confluens*. Acta Soc. Sci. Fennicae, xiv, p. 337.

1900 HARPER, R. A. Sexual Reproduction in *Pyronema confluens*, and The Morphology of the Ascocarp. Ann. Bot. xiv, p. 321.

1905 CLAUSSEN, P. Zur Entwickelungsgeschichte der Ascomyceten. *Boudiera*. Bot. Zeit. lxiii, p. 1.

1907 DANGEARD, P. A. Recherches sur le développement du périthèce chez les Ascomycètes. Le Botaniste, x, pp. 247 and 259.

1909 BROWN, W. H. Nuclear Phenomena in *Pyronema confluens*. Johns Hopkins Univ. Circ. vi, p. 42.

1912 CLAUSSEN, P. Zur Entwickelungsgeschichte der Ascomyceten. *Pyronema confluens*. Zeitschr. f. Botanik, iv, p. 1.

1915 BROWN, W. H. The Development of *Pyronema confluens*, var. *inigneum*. Am. Journ. Bot. ii, p. 289.

### *Pezizaceae*

The Pezizaceae form a rather large group. The ascocarp is superficial, sessile or stalked, usually with a well-marked peridium fleshy or waxy in consistency, and soon decaying after maturity. The spores are usually hyaline and continuous (though septate in some small species) and are typically uniseriate. The asci do not project above the level of the disc at maturity, as they do in the Ascobolaceae. The species are mostly saprophytic, many occurring on the ground, and a few, especially the smaller forms, on dung. The subdivisions depend on the shape of the spores, the size and consistency of the ascocarp, and the presence or absence of hairs.

In the majority of forms the fruit is fleshy and without hairs; these species are often grouped together in the single genus *Peziza*, but it is probably more convenient to separate them. The name *Peziza* is retained for large species with a sessile or subsessile cup, regular in form and two centimetres or more across as in *P. vesiculosa*. The genus *Humaria* includes similar but smaller species, often less than one centimetre in diameter. In *Otidea* the sides of the ascophore are laterally split, or vertically incurved and wavy. In *Acetabula* and *Geopyxis* the ascophore is stalked. In *Lachnea*, as well as in some other genera, the fruit is beset with hairs and in *Sepultaria* it is hairy and more or less sunk in the soil.

*Lachnea stercorea* is a small orange species occurring during the winter and spring on the dung of various animals, especially of cows. With *Humaria granulata* and *Ascobolus furfuraceus*, it is among the very common coprophilous forms, appearing in many parts of Britain with great regularity when the *Piloboli* have died down, and the cow pad is beginning to dry. It is about 4 mm. in diameter and is furnished with numerous stout, septate hairs.

The archicarp arises as a side branch from the vegetative mycelium, and divides to form four or more cells. The terminal cell or oogonium is oval in shape and larger than the others. It contains between two and three hundred nuclei and is filled with finely granular cytoplasm. In the cell next below the oogonium, the cytoplasm is also more dense and the nuclei more numerous than in the other cells of the fertile branch.

Hyphae grow up from the lower cells of the archicarp, and from the branch which bears it, and form a dense weft above which the oogonium rises.

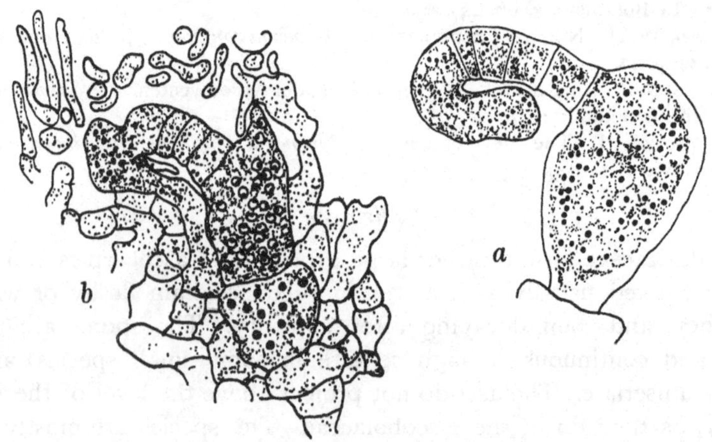


Fig. 65. *Lachnea stercorea* (Pers.) Gill.; *a*. young archicarp, × 800; *b*. archicarp and antheridium, × 500; P. Highley del.

The oogonium sends out either laterally, or from its apex, a stout branch or trichogyne. It is cut off by a wall and divides into four to six cells, the terminal of which is much larger than the others (fig. 65*a*). The tip of the trichogyne protrudes for a time beyond the developing sheath, but later, with the whole fertile branch, it is enclosed by vegetative hyphae.

At this stage a large, more or less oval sac is often found to be continuous with the terminal, or receptive, cell of the trichogyne into which a proportion of its contents pass (fig. 65 *b*). There seems no doubt that this sac is the antheridium, but its development is not known, and there is no evidence that its contents ever pass beyond the receptive cell of the trichogyne. Indeed all the available evidence shows that both antheridium and trichogyne are now merely vestigial structures.

Nevertheless the development of the oogonium continues, its nuclei increase to something over 500 in number, and ascogenous hyphae bud out. Before passing into these the oogonial nuclei fuse in pairs, so that normal fertilization is here replaced by the union of female nuclei. The ascogenous hyphae branch and give rise to asci, in each of which eight spores are produced in the usual way. The karyokinetic figures are small but very clear, there are four chromosomes in the first and second divisions, but in the third telophase only two have been recorded. There is some evidence that the chromosomes show regular and characteristic differences of form, which reappear in successive divisions.

The peridium, though much better developed than in the Pyronemaceae, is never completely closed, as in *Humaria* or *Ascobolus*, across the top of the ascocarp. The paraphyses are numerous and contain orange granules.

*Lachnea scutellata* occurs on decaying wood, forming bright red apothecia. The archicarp consists of seven to nine cells, the subterminal of which enlarges to form the oogonium. The nuclei in this cell divide, and, according to Brown, show five short, stout chromosomes. He did not observe nuclear fusion or association in the oogonium, but regards the nuclei lying in contact as the two daughter nuclei of a single mitosis. Large ascogenous hyphae develop, undergo septation, and branch freely. Their tips bend over and asci are formed in the usual way from the penultimate cells. The terminal cells may undergo further growth and give rise, as in several other Discomycetes, to new asci. Nuclear fusion takes place in the young ascus, and is followed by a meiotic reduction. Five gemini are recorded, but, in the anaphase of the first division, ten chromosomes travel towards each pole. This Brown takes to indicate an early fission of the daughter chromosomes. In the second and third divisions five chromosomes are seen throughout. Brown infers the occurrence of a single fusion in this species, that in the ascus, and a single reducing division[1].

*Lachnea cretea* has a pale buff apothecium, beset with hairs (fig. 66*a*). It has been found on plaster ceilings, and, like many other saprophytic species, grows readily in artificial culture.

[1] The magnification of Brown's figures of the divisions in the oogonium is enormous (× 11,200), and their details should therefore probably be received with some caution.

The archicarp (figs. 66 *b–e*) consists of a long, branched, multicellular trichogyne, an oogonial region of three or four coenocytic cells, and a multicellular stalk. No antheridium has been observed. In the trichogyne (fig. 66 *e*), pores are formed between the adjacent cells, and are closed after a time by "callus" pads. In the central part of the archicarp the transverse septa are completely broken down, so that a very wide passage is formed, and nuclei pass readily from cell to cell (fig. 66 *f*). All the cells give rise

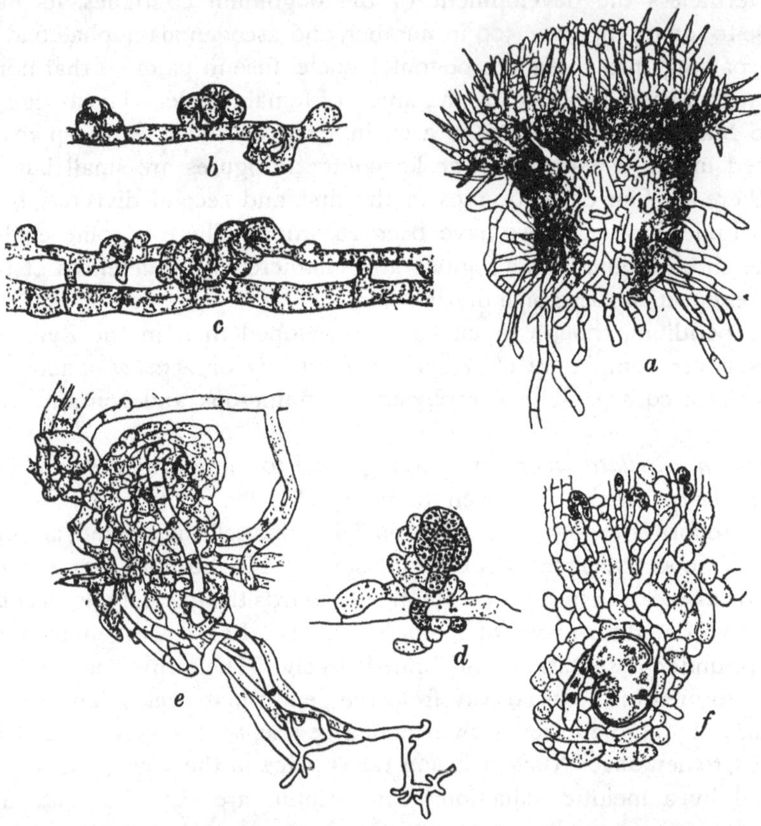


Fig. 66. *Lachnea cretea* Phil.; *a*. mature ascocarp, ×90; *b*. *c*. development of archicarp, ×300; *d*. older archicarp showing crowded nuclei, ×400; *e*. mature archicarp with elaborately branched trichogyne, ×400; *f*. three ascogonial cells united by very large pores, ×400.

to ascogenous hyphae. Thus the oogonial region, though developmentally multicellular, is for all practical purposes unicellular at maturity, and offers no greater difficulties in the way of fertilization than the oogonium of *Pyronema* itself.

The branched character of the trichogyne is exceptional among Discomycetes; it might, no doubt, facilitate the establishment of contact with an attached antheridium if the latter developed at a distance. But branching might also be regarded as a secondary or vegetative development, appearing after normal fertilization had ceased to occur.

The presence of pores in the transverse septa of the trichogyne suggests that the function of that organ in relation to an antheridium has only recently been lost.

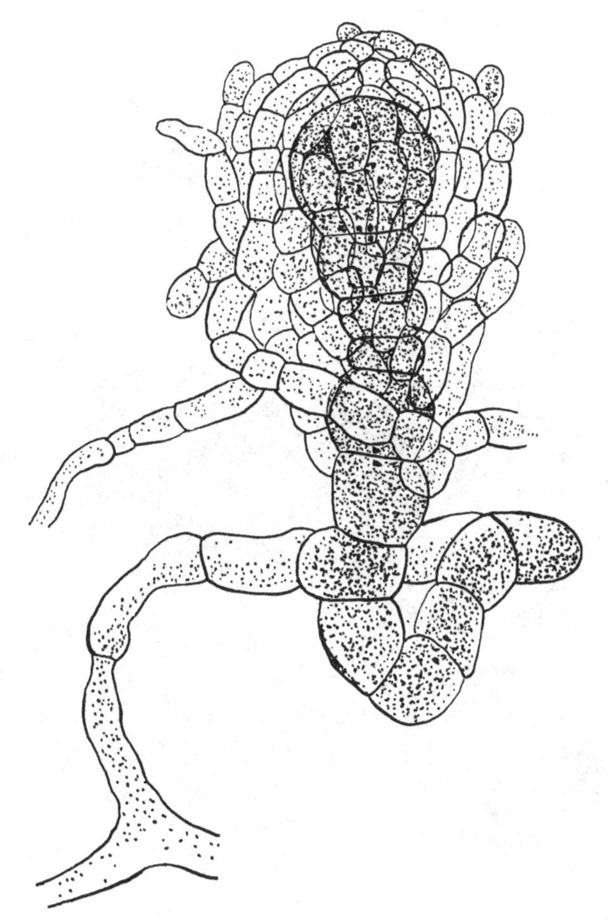

Fig. 67. *Humaria granulata* Quel.; young archicarp, × 320; after Blackman and Fraser.

The ascogenous hyphae contain many nuclei irregularly arranged. Asci are formed in the usual way; their nuclei show about eight chromosomes in the first division. Owing to the small size of the nuclei further cytological details have not been studied in this species.

*Humaria granulata* is a common red or orange coprophilous form. The archicarp develops as a side branch from an ordinary hypha. The apical cell of this branch increases in size and becomes spherical, forming the oogonium (fig. 67); it contains large numbers of well-marked nuclei. When it is full grown the oogonial nuclei fuse in pairs (fig. 68 *a*), and the fusion nuclei pass into the ascogenous hyphae (fig. 68 *b*). There is no sign of either trichogyne or antheridium.

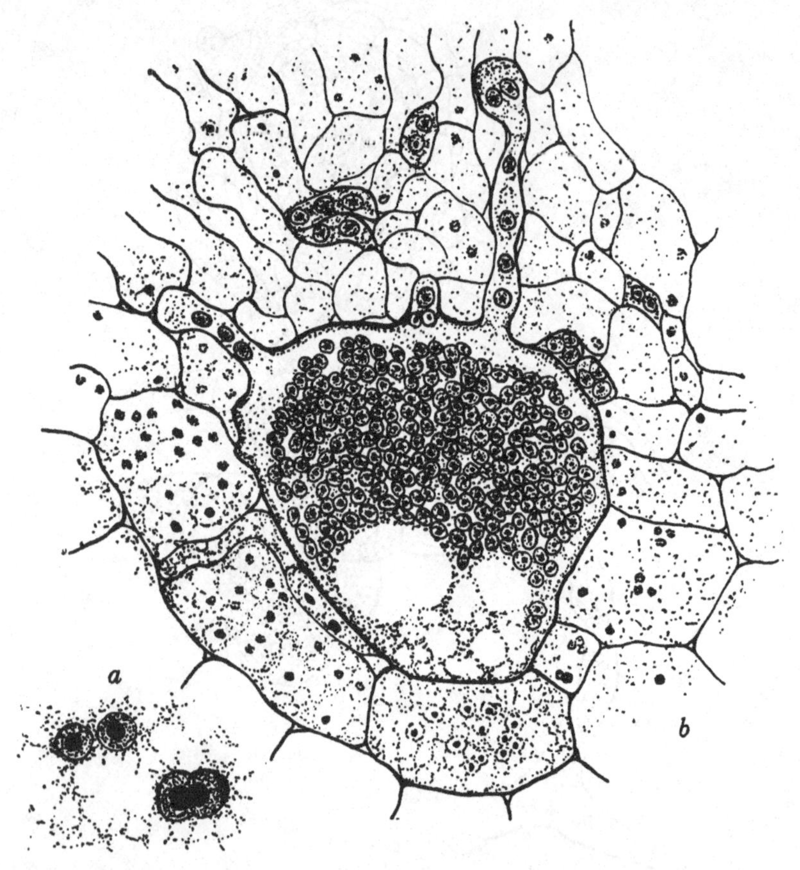


Fig. 68. *Humaria granulata* Quel.; *a*. fusion of nuclei in oogonium, × 3200; *b*. oogonium giving rise to ascogenous hyphae, × 1250; after Blackman and Fraser.

Vegetative cells grow up and invest the archicarp, forming a close pseudoparenchymatous sheath in which the ascogenous hyphae ramify. They give rise at last to asci in the usual way.

Four chromosomes have been recorded in the ascogenous hyphae, eight in the first division in the ascus and four in the two subsequent mitoses. This implies that the gametophytic number is four, and that the gemini are formed immediately after the fusion in the oogonium, so that in the ascogenous hyphae there are four bivalent instead of eight univalent chromosomes. In the meiotic prophase which follows the fusion in the ascus, there is a double number of gemini, since two sporophytic nuclei have united.

In *Humaria granulata*, the antheridium has disappeared and normal fertilization is replaced by fusion of female nuclei in pairs in the oogonium.

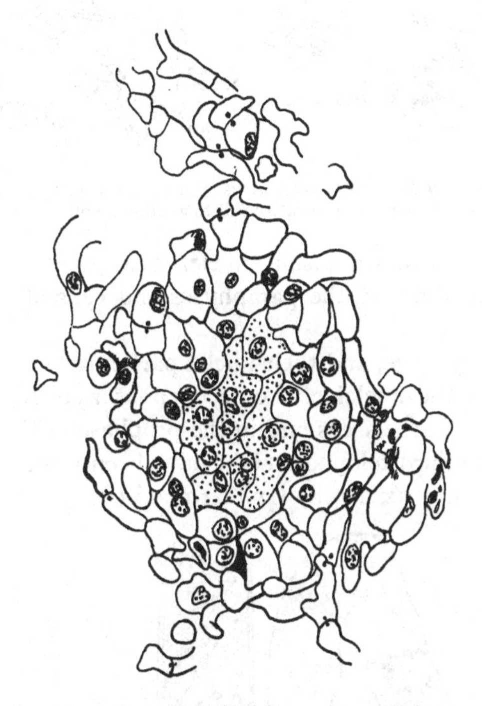

Fig. 69. *Humaria rutilans* (Fr.) Sacc.; very young ascocarp, × 500.

In another species of this genus, *Humaria rutilans*[1], reduction has gone yet further and not even an archicarp is produced. The apothecium arises as a dense weft of tangled filaments, which for a time differ from one another only in the relatively thick walls of the outer hyphae, and the richer protoplasmic content of the inner (fig. 69). Each cell of the weft contains one or a few nuclei. After a while the nuclei in the central part of the mass may be seen to be of two sizes, and the smaller have been found to fuse in

[1] *Humaria rutilans* (Fr.) Sacc. = *Peziza rutilans* Fr. in Boudier, *Icones*, Pl. 315.

pairs (fig. 70 *a*), giving rise to the larger. Sometimes in connection with this process a nucleus migrates through the wall from one cell to another (fig. 70 *b*), as in prothallia of ferns. Thus in *H. rutilans*, where the sexual organs are completely lacking, normal fertilization is replaced by the union of vegetative nuclei in pairs.

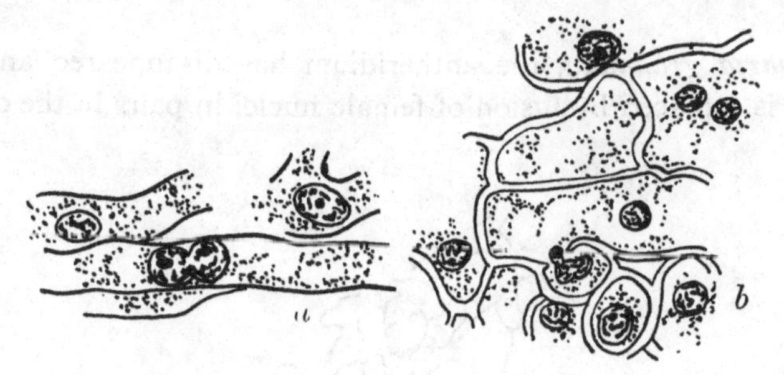


Fig. 70. *Humaria rutilans* (Fr.) Sacc.; *a.* fusion in a vegetative hypha; *b.* migration of nucleus from one vegetative cell to another; both × 1100.

The cells which contain fusion nuclei now give rise to ascogenous hyphae, while, from the rest, the paraphyses and cells of the outer sheath arise.

The asci are very large, and their nuclei particularly clear. The number of chromosomes in the nuclei of the ascogenous hyphae, and in the first and second divisions in the ascus and in the prophase of the third is sixteen (figs. 71, 72). In the third telophase eight have been recorded by Maire and by Fraser (fig. 73), and sixteen by Guilliermond (fig. 74).

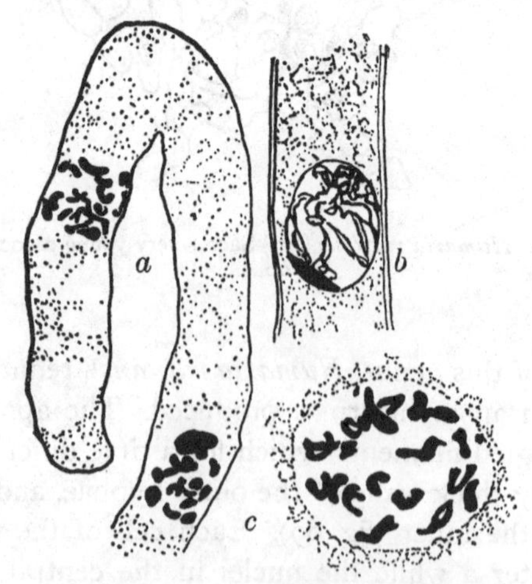


Fig. 71. *Humaria rutilans* (Fr.) Sacc.; *a.* ascogenous hypha showing sixteen chromosomes in each nucleus, × 1950; *b.* fusion nucleus of ascus passing out of synapsis, × 1300; *c.* fusion nucleus of ascus showing sixteen gemini, × 1950.

In several other members of the Pezizaceae, for example in *Peziza vesiculosa* (Fraser and Welsford) and *Peziza tectoria*, development apparently takes place, as in *Humaria rutilans*, without the formation of sexual organs.

In *Otidea aurantia* (Fraser and Welsford), a large cell, no doubt part of an archicarp, has been recorded in the early stages, and in *Peziza theleboloides*, *Humaria Roumegueri*, and *H. carbonigena*, there is a well-marked oogonial region of one or more cells.

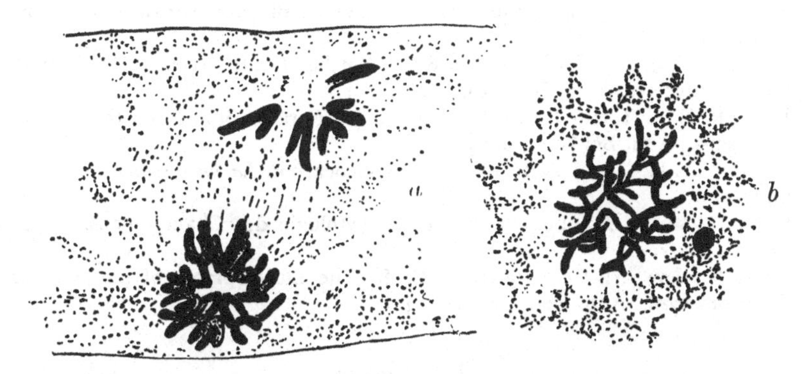


Fig. 72. *Humaria rutilans* (Fr.) Sacc.; *a.* telophase of second division in ascus, × 3370; *b.* prophase of third division in ascus, showing sixteen curved chromosomes, × 2808.

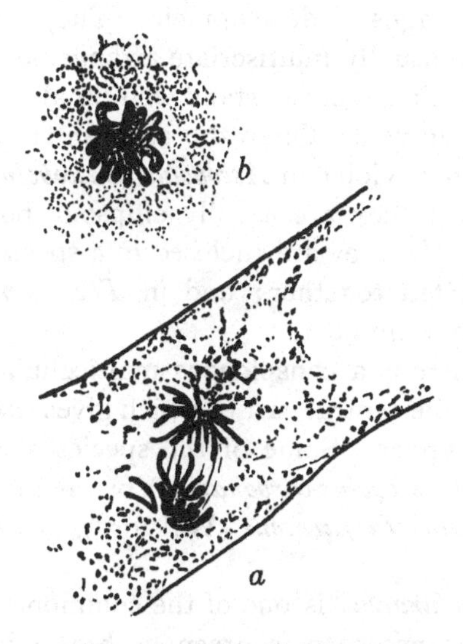


Fig. 73. *Humaria rutilans* (Fr.) Sacc.; *a.* metaphase of third division in ascus, × 2080; *b.* polar view of telophase of third division in ascus, showing eight curved chromosomes, × 3100.

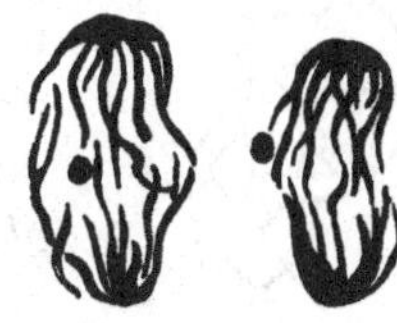

Fig. 74. *Humaria rutilans*; telophase of third division in ascus; after Guilliermond.

## PEZIZACEAE: BIBLIOGRAPHY

1905 GUILLIERMOND, A. Remarques sur le Karyokinèse des Ascomycètes. Ann. Myc. iii, p. 343.
1905 MAIRE, R. Recherches cytologiques sur quelques Ascomycètes. Ann. Myc. iii, p. 123.
1906 BLACKMAN, V. H. and FRASER, H. C. I. On the Sexuality and Development of the Ascocarp in *Humaria granulata.* Proc. Roy. Soc. B. 77, p. 354.
1907 FRASER, H. C. I. On the Sexuality and Development of the Ascocarp in *Lachnea stercorea.* Ann. Bot. xxi, p. 349.
1908 FRASER, H. C. I. Contributions to the Cytology of *Humaria rutilans.* Ann. Bot. xxii, p. 35
1908 FRASER, H. C. I. and WELSFORD, E. J. Further Contributions to the Cytology of the Ascomycetes. Ann. Bot. xxii, p. 465.
1909 FRASER, H. C. I. and BROOKS, W. E. St J. Further Studies on the Cytology of the Ascus. Ann. Bot. xxiii, p. 537.
1911 BROWN, W. H. The Development of the Ascocarp in *Lachnea scutellata.* Bot. Gaz. lii, p. 275.
1911 GUILLIERMOND, A. Les Progrès de la cytologie des Champignons. Prog. Rei Bot. vi, p. 389.
1913 FRASER (GWYNNE-VAUGHAN), H. C. I. The Development of the Ascocarp in *Lachnea cretea.* Ann. Bot. xxvii, p. 554.

### *Ascobolaceae*

The large majority of the Ascobolaceae are coprophilous; their ascocarp is soft and fleshy or somewhat gelatinous, and they possess a well-marked sheath which is closed during the early stages of development. They are distinguished from the Pezizaceae by the usually multiseriate arrangement of their spores, and by the fact that, when ripe, the asci stand well up above the hymenium before the spores are discharged. Often the asci are large and few in number; the spores are brown or violet in *Ascobolus*, *Saccobolus* and *Boudiera*, hyaline in the other genera; they are usually ellipsoid, but round in *Boudiera* and *Cubonia*; in *Saccobolus* they are enclosed in a special membrane within the ascus and are ejected together; and in *Thelebolus* and *Rhyparobus* they are sixteen or more in number.

In most of the species investigated there is a conspicuous multicellular coiled archicarp, the central part of which gives rise to ascogenous hyphae. Some of the species also produce conidia (*Ascobolus carbonarius*), or chlamydospores (*Ascobolus furfuraceus* (Welsford), *Ascophanus carneus*).

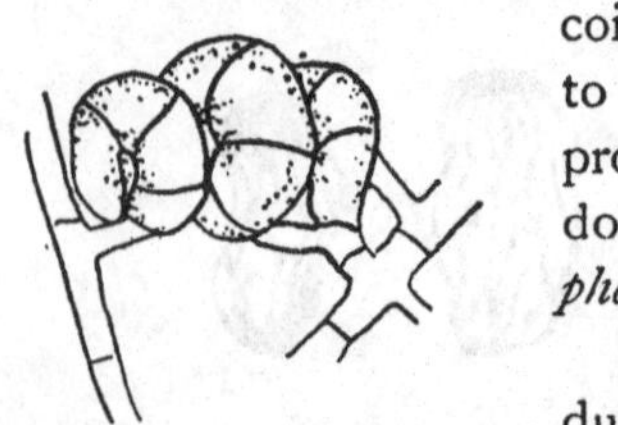

Fig. 75. *Ascobolus furfuraceus* Pers.; archicarp, × 740; after Dodge.

*Ascobolus furfuraceus* is one of the commonest dung species, the ascocarp is green or brown in colour with a characteristic scurfy margin. The archicarp (fig. 75) consists of sometimes as many as twenty (Dodge), sometimes a much smaller number of cells. These are at first uninucleate (Harper, Welsford), or multinucleate (Dangeard); later they always contain numerous nuclei (fig. 76 *a*). One of them, usually the

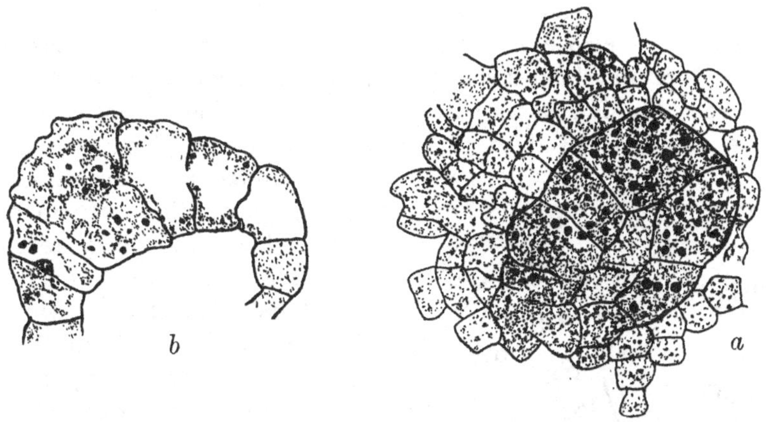


Fig. 76. *Ascobolus furfuraceus* Pers.; *a*. young archicarp, × 750; *b*. rather older specimen showing pores between the cells, × 625; after Welsford.

fourth from the apex (Welsford), enlarges, buds out ascogenous hyphae and functions as the oogonium. Those near the base form a stalk, and those towards the apex may be regarded as constituting a now functionless trichogyne.

The cells on each side of the oogonium communicate with it by means of pores (fig. 76 *b*). Additional nuclei pass into it from both the stalk and terminal cells, and Welsford has observed their fusion in pairs in the oogonium. The fusion nuclei pass into the ascogenous hyphae. The asci are large and produce each eight spores which are violet or brownish in colour; the epispore is characteristically sculptured at maturity. There are eight chromosomes in the first division in the ascus, and four in the second and third (Dangeard (fig. 13), Fraser and Brooks).

In *Ascobolus glaber* the archicarp is larger and more twisted than in *A. furfuraceus*, and consists of some twenty or thirty cells from one or more of which the ascogenous hyphae develop (Dangeard).

In *Ascobolus Winteri*, a form occurring on goose dung, and apparently limited to this habitat, the archicarp (fig. 77), as described by Dodge, consists of three parts, a stalk of two or three cells, a series of larger, central cells, which give rise to the ascogenous hyphae, and a terminal row of three

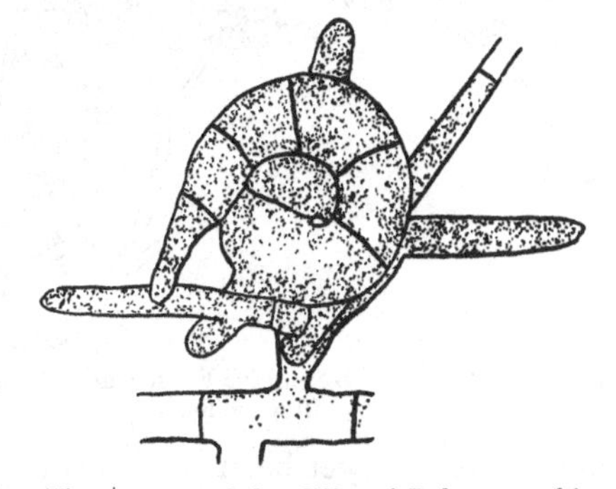

Fig. 77. *Ascobolus Winteri* Rehm.; archicarp, × 1080; after Dodge.

or four cells, which diminish gradually in diameter and which he terms a trichogyne.

In *Ascobolus immersus* the mycelium consists of multinucleate cells, the archicarp is larger than that of *A. Winteri* and contains some twenty divisions, it is otherwise very similar. The cells contain numerous large nuclei and pores develop between them; the ascogenous hyphae arise from a single cell. Ramlow observed nuclear fusions in the central cell of the archicarp, but referred them to bad fixation. His explanation may be adequate here, but it does not invalidate the observations of authors who have recorded fusions in properly fixed material. In archicarps which he held to be satisfactorily fixed Ramlow saw pairing of nuclei in the ascogenous cell, and records that they wandered in pairs into the ascogenous hyphae (fig. 78), not fusing till the asci were about to develop. In the divisions of the ascus the number of chromosomes is stated by Ramlow to be sixteen throughout, but he does not figure the essential third anaphase.

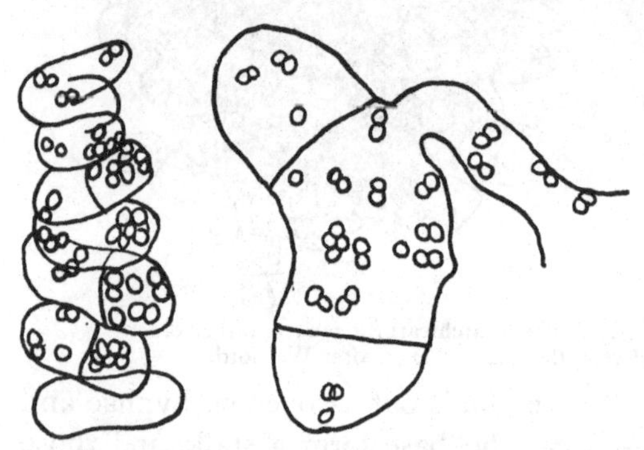

Fig. 78. *Ascobolus immersus* Pers.; archicarps showing paired nuclei, × 1000; after Ramlow.

*Ascobolus carbonarius* occurs on burnt ground among charcoal. The ascocarp is scurfy (furfuraceous), and greenish, or later brownish in colour. Numerous conidia are formed on the mycelium, and, according to Dodge, it is from a conidium, germinating while still attached to its stalk, that the archicarp is produced. It consists of a multicellular stalk, a fertile portion which contains twenty to forty cells arranged in a loose irregular spiral, and a terminal trichogyne more or less coiled, tapering towards the end, and including some ten to twenty cells. The apex of this trichogyne is found to wrap itself tightly round a second conidium, attached, like the first, to its stalk (fig. 79). This conidium is regarded by Dodge as the antheridium, but no cytological details have as yet been published.

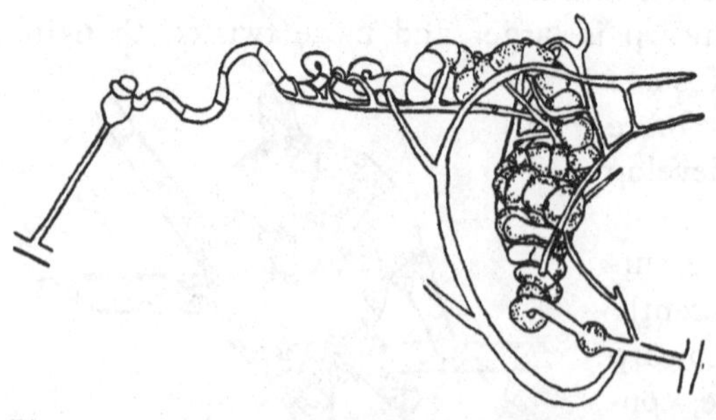

Fig. 79. *Ascobolus carbonarius* Karst.; archicarp, × 280; after Dodge.

*Ascophanus carneus* is a somewhat variable species. Its red, pink, or orange apothecia occur on the dung of cows and rabbits, on old leather, rope and similar habitats. Chlamydospores are sometimes produced.

As in *Ascobolus*, the archicarp is a coiled, multicellular filament; it varies considerably both in the size and number of its cells and in the amount of twisting which it undergoes.

The central oogonial region includes three to seven large cells with granular contents. Between this and the parent hypha is a stalk of variable length and beyond it is a terminal portion (or trichogyne) of not more than seven cells which are narrower than the rest and appear to degenerate early (Cutting).

The cells come into communication with one another by large pores (fig. 81 *a*), and Cutting has shown that nuclear fusions (fig. 81 *b*) take place in all the cells of the oogonial region, and that all of them give rise to ascogenous hyphae.

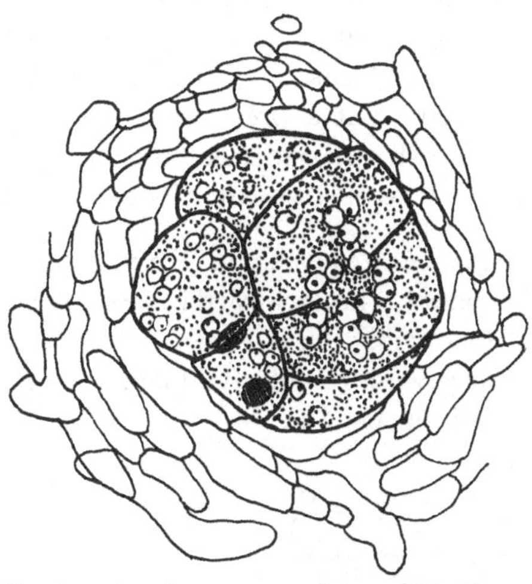

Fig. 80. *Ascophanus carneus* Pers.; old archicarp, showing associated nuclei, × 800; after Ramlow.

Ramlow also saw nuclear fusions in these cells, but he explained them, as in *Ascobolus immersus*, as due to bad fixation. He also saw and figured exceptionally large nuclei, which had apparently become swollen, and were about to degenerate. For him the normal process is the association of nuclei in pairs (fig. 80) without fusion, and their passage, still associated, into the ascogenous hyphae; here walls are formed so that the hypha consists of a series of binucleate cells. These

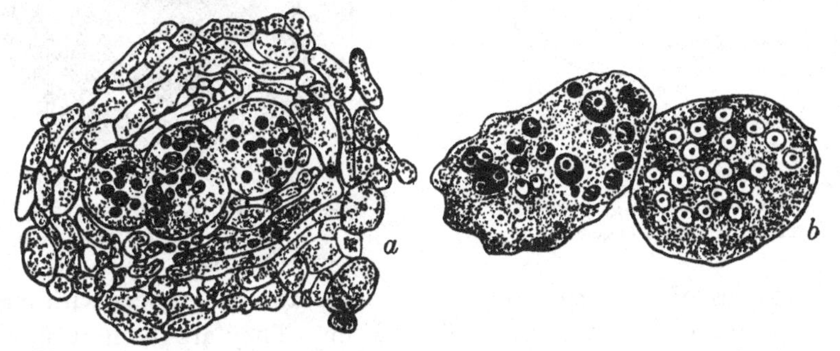


Fig. 81. *Ascophanus carneus* Pers.; *a*. section through young ascocarp, showing nuclear fusion in two cells of the archicarp, × 580; *b*. two cells of an archicarp, showing nuclear fusions, × 1240; after Cutting.

nuclei, when satisfactorily fixed, showed a well-marked centrosome. Ramlow was unable to see whether one or several cells of the archicarp gave rise to ascogenous hyphae; an investigation sufficiently searching to determine this point might have led to the recognition of nuclear fusions in normal material.

The ascospore has a single large nucleus, and gives rise to multinucleate germ-tubes in which Ramlow's figures show numerous nuclei in pairs (fig. 16).

In *Ascophanus ochraceus* Dangeard describes eight to fifteen oogonia as taking part in the formation of a single fruit. These, it would appear, are all borne upon the same hypha; they may arise from adjacent cells, and indeed sometimes open into one another, so that the whole series seems equivalent to the oogonial region of *A. carneus*. Each cell, however, is described as bearing a twisted, multicellular trichogyne, which would indicate that each is an independent organ. Large ascogenous hyphae arise from the several "oogonia," and the view suggests itself that the so-called trichogyne may be in fact a premature ascogenous hypha. It is, at least, difficult to distinguish the one from the other in Dangeard's figures, and the species certainly requires further investigation.

*Saccobolus violascens* is a violet or greyish violet species about 1 mm. in diameter. The archicarp is a coiled structure and is divided into only three or four cells (Dangeard), the central of which gives rise to ascogenous hyphae, while vegetative filaments grow up from the stalk and neighbouring mycelium (fig. 82).

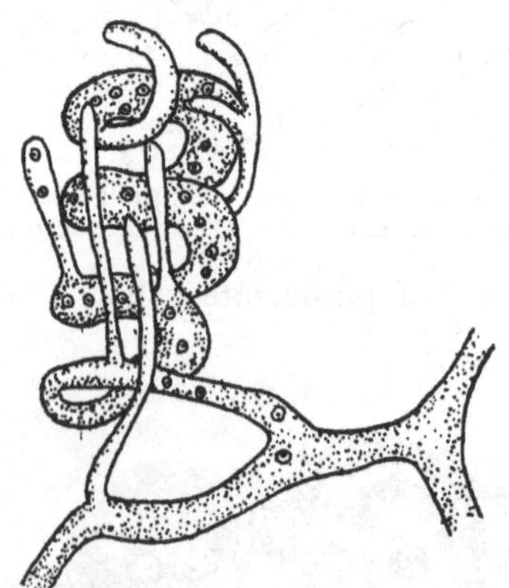

Fig. 82. *Saccobolus violascens* Boud.; archicarp; after Dangeard.

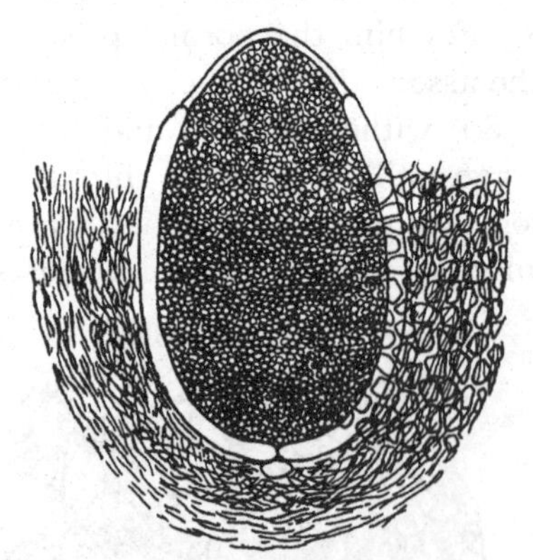

Fig. 83. *Thelebolus stercoreus* Tde.; ascocarp with single ascus, × 250; after Brefeld.

The species of *Rhyparobius* and *Thelebolus*, the two genera with many-spored asci, are all minute, coprophilous forms. They are distinguished by the fact that *Rhyparobius* produces several large asci, and *Thelebolus* only one (fig. 83). In both genera the cells of the mycelium are uninucleate.

In *Rhyparobius* (*Thecotheus*) *Pelletieri* Overton has described several multicellular archicarps, each rather like the single scolecite of *Ascobolus*. The cells are not connected by pores, and ascogenous hyphae arise from several in each archicarp.

In *R. brunneus* Dangeard reports a single archicarp, consisting of a short, somewhat twisted branch. Ramlow has also recorded a single archicarp in *R. polysporus* and Barker in an unnamed species. Overton has made some study of the development of the numerous spores in *R. Pelletieri*. The ascus nucleus divides as usual to form eight free nuclei, these undergo a period of rest and growth and then divide further till thirty-two free nuclei are formed. Around these the spores are delimited in the usual way.

*Thelebolus stercoreus* has a mycelium of uninucleate cells, from one of which the archicarp arises as a thick branch containing a single nucleus. Later two, four, and finally eight, are seen (fig. 85), and then septation takes

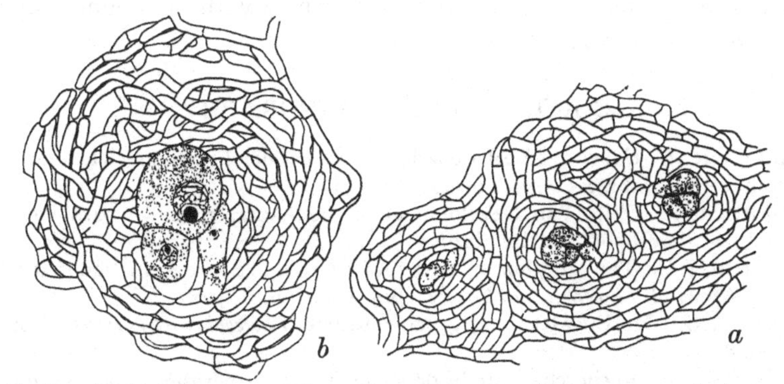


Fig. 84. *Thelebolus stercoreus* Tde.; *a*. young ascocarp with binucleate asci; *b*. ascus containing fusion nucleus, both × 810; after Ramlow.

place, so that a row of cells is formed. Most of these are uninucleate, but one contains two nuclei (fig. 84*a*); it enlarges and becomes the single ascus; in it the two nuclei fuse (fig. 84*b*). The definitive nucleus divides karyokinetically, sometimes as many as ten times, so that 1042 nuclei are formed.

Fig. 85. *Thelebolus stercoreus*, Tde.; development of archicarp, × 1750; after Ramlow.

Spore-formation takes place apparently in the usual way. The wall of the ripe ascus is about $2\mu$ thick, but a thinner region is present at the apex, so that a concave papilla is differentiated, which is concerned in the dehiscence of the ascus. A sheath of vegetative hyphae grows up from the surrounding cells.

In *Thelebolus Zukalii* the origin of the ascocarp has been observed from a pair of intertwined hyphae (Ramlow, 1914), but the cytology and further development have not yet been described.

The closest analogy to the development of the ascocarp in *Thelebolus* is perhaps to be found in *Sphaerotheca* among the Erysiphales. In both we have a hypha which is at first uninucleate, later multinucleate. In both it divides to form a row of cells most of which enclose one nucleus. In both a single binucleate cell, typically penultimate, gives rise to the single ascus in which nuclear fusion takes place. But in *Sphaerotheca* the original uninucleate structure is the fertilized oogonium, while in *Thelebolus stercoreus* an antheridium has not been demonstrated. It remains to be seen what light the investigation of *Th. Zukalii* will throw on these homologies. In view of the direct transformation of one cell of the row into an ascus, it becomes unjustifiable to correlate the septate structure here with the young archicarp of the *Ascoboli* or *Ascophani*.

## ASCOBOLACEAE: BIBLIOGRAPHY

1896 HARPER, R. A. Ueber das Verhalten der Kerne bei der Fruchtentwickelung einiger Ascomyceten. Jahr. für wiss. Bot. xxix, p. 655

1903, 4 BARKER, P. T. B. The Development of the Ascocarp in Rhyparobius. Rept. Brit. Assoc. Southport and Cambridge.

1906 OVERTON, J. B. The Morphology of the Ascocarp and Spore-formation in the many spored Ascus of *Thecotheus Pelletieri*. Bot. Gaz. lxii, p. 450.

1906 RAMLOW, G. Zur Entwickelungsgeschichte von *Thelebolus stercoreus*. Bot. Zeit. lxiv, p. 85.

1907 DANGEARD, P. Recherches sur le développement du périthèce chez les Ascomycètes. Le Botaniste, x, p. 304.

1907 WELSFORD, E. J. Fertilization in *Ascobolus furfuraceus*. New Phyt. vi, p. 156.

1909 CUTTING, E. M. On the Sexuality and Development of the Ascocarp in *Ascophanus carneus*. Ann. Bot. xxiii, p. 399.

1909 FRASER, H. C. I. and BROOKS, W. E. St J. Furthur Studies on the Cytology of the Ascus. Ann Bot. xxiii, p. 537.

1912 DODGE, B. O. Methods of Culture and the Morphology of the Archicarp in certain Species of the Ascobolaceae. Bull. Torrey Bot. Club, xxx, p. 139.

1914 RAMLOW, G. Beiträge zur Entwickelungsgeschichte der Ascoboleen. Myc. Centralbl. v, p. 177.

### *Helotiaceae and Mollisiaceae*

The Helotiaceae and Mollisiaceae are distinguished from the Pezizaceae by the fact that their peridium differs more or less definitely in structure from the hypothecium. In Helotiaceae the peridium is prosenchymatous, composed of elongated parallel hyphae, usually light in colour and thin-walled. In the Mollisiaceae it is parenchymatous, of round or polygonal cells usually thick-walled and dark-coloured. In both families the ascus opens by the ejection of a plug, and not, as in most Discomycetes, by a lid.

The apothecia of the majority of forms included in these two families are small, often stalked, sometimes attached to a sclerotium; they are waxy in consistency and may be glabrous or hairy. Most are saprophytes, often occurring on dead plants, some are parasitic. In Helotiaceae they are almost always sunk in the substratum (immersed), and in Mollisiaceae frequently superficial.

Among the Mollisiaceae *Pseudopeziza Trifolii* is parasitic and causes the leaf-spot disease of clover. The leaves become increasingly spotted and die, so that the crop is often seriously injured. In this case the ascocarps are sessile and distinctly erumpent, developing within the tissue of the leaf, and breaking through the epidermis at maturity. There are several other species of *Pseudopeziza*, most on dead stems and leaves, a few on the living tissues of wild plants. The species of *Tapesia* occur on wood, branches, and dead leaves. The ascocarps are stipitate and pilose or downy, they are found in groups seated on a spreading weft of branched interwoven hyphae, by means of which the genus is readily distinguished. *T. fusca* is to be found on fallen twigs of larch and other plants.

Among the Heliotiaceae the genus *Helotium* includes a number of species found on dead leaves, stems, beechmast, and similar habitats; these fungi are light-coloured, waxy and frequently stipitate.

Another large genus, *Dasyscypha*, has a sessile or short-stalked ascocarp, thin and delicate in texture, and externally pilose; the species are saprophytic or parasitic. *Dasyscypha Willkommii* is the cause of a serious disease, the well-known Larch Canker. The apothecia are externally yellow with an orange disc. The ascospores give rise to germ-tubes which are unable to penetrate the bark, but obtain entrance through wounds caused by hail, ice or snow, or by the destruction of the needles by insects. The Larch moth (*Coleophora laxicella*), for instance, is known to cause less injury in mountainous than in lower regions, and the fungal disease is proportionately less prevalent in the mountains. The mycelium ramifies chiefly through the soft bast, but may penetrate the wood as far as the pith. It spreads only in the autumn and winter, never in summer when the growth of the host is active. Where it spreads into the bark the tissues turn brown and shrivel, causing depressed canker spots in which yellowish white stromata are produced. These give rise to minute unicellular conidia, and later, if the atmosphere is sufficiently moist, to ascocarps.

In the genus *Sclerotinia* the stalked ascocarps arise from sclerotia (fig. 86). A number of species are parasitic: *S. tuberosa* on *Anemone nodosa*; *S. sclerotiorum* on the potato, cabbage and other hosts in the stems of which the sclerotia are formed; *S. fructigena* and *S. cinerea* on species of *Prunus* and *Pyrus* where they give rise to brown rot, blossom wilt and other pathological conditions; *S. bulborum* on hyacinth and other bulbs, and various species on members of the Vaccinieae, where the sclerotia are formed on the fruits. In *S. Vaccinii* the conidia are produced in chains and are separated by small cellulose disjunctors. They have a characteristic smell of almonds and are carried to the flower by insects, and probably also by wind; they germinate to form septate hyphae which enter and fill the ovary. The

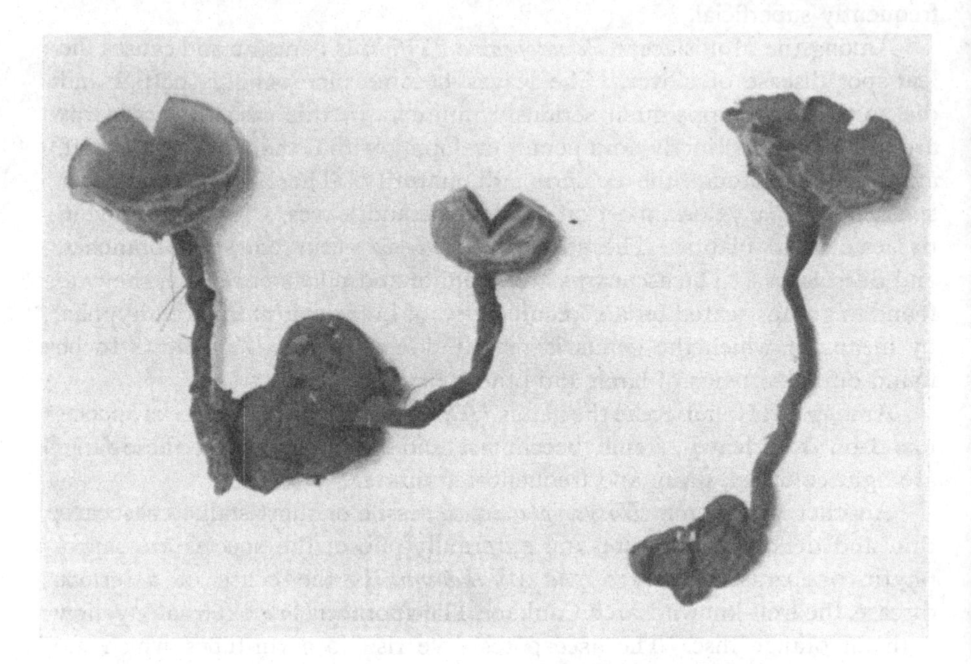

Fig. 86. *Sclerotinia tuberosa* (Hedw.) Fuck.; sclerotia and apothecia, nat. size.

mummified berries fall prematurely, lie during winter on the earth, and in spring give rise to the goblet shaped apothecia. In other species the conidia are borne on a conidiophore and belong to the form-genus *Botrytis*; the conidial phase on *Prunus* and *Pyrus* is known as *Monilia*. The ascospores are unicellular and hyaline and often of unequal size.

### *Celidiaceae, Patellariaceae, and Cenangiaceae*

In the previously described families the consistency of the ascocarp is either fleshy or waxy. In the following three, Celidiaceae, Patellariaceae, and Cenangiaceae, it is leathery, horny, or cartilaginous, and the ends of the paraphyses are interwoven to form a layer above the asci known as the **epithecium**. The hypothecium is well developed, the ascospores are sometimes more than eight in number and are one to many-celled; in some species pycnidia are present. The three families are sometimes grouped together as Dermateaceae.

Some of the Celidiaceae occur on wood or bark, but the majority are parasitic on the thalli or apothecia of lichens, their hyphae ramify among the living tissues of the host, and they were at first believed to be themselves lichen species. They are, however, without a thallus, and so without an algal constituent, and the host plant is clearly distinguished by its own fructification developed in the absence of the parasite. The fruits of *Celidium varians*, for example, form black points on the apothecia, or rarely on the thallus of the lichen *Lecanora glaucoma*. In this family the peridium is absent or but little developed.

The Patellariaceae are for the most part saprophytic, but include also a number of lichen parasites. These are erumpent, but the saprophytic forms are superficial, and are thus differentiated from the Cenangiaceae. They are distinguished from the Celidiaceae by the well-marked peridium of small, dark-coloured cells. The fruits are closed at first and either become flattened out as they develop, or open by a narrow or a star-shaped slit.

In the Cenangiaceae the ascocarps are erumpent, sometimes developed on a stroma. They are dark-coloured, with a tough or somewhat gelatinous sheath, and, when mature, are cup or pitcher shaped; pycnidia or spermogonia are present in some genera.

*Bulgaria polymorpha*, one of the best known species, occurs on dead trunks of trees, particularly beech. The cup is 1 to 4 cm. across, and is externally umber brown. The hymenium is black and shining and level or almost level with the top of the cup. The ascocarps burst through the bark as small, rusty brown, scurfy knobs, which gradually expand at the apex. The substance is soft and tough, resembling india-rubber in consistency and appearance. The species is readily distinguished by its four, slightly curved, brown ascospores. It is stated to be a dangerous enemy of the oak, but details of its parasitism are not known.

The genus *Coryne* is placed by many systematists in the neighbourhood of *Bulgaria*. *C. sarcoides* is a common species on rotten trunks and stumps. The apothecia are crowded and dull red or purple in colour. Amongst them, or often occurring alone, are the conidial fructifications, rather paler in colour. Minute conidia are abstricted from the ends of the fertile hyphae. The ascospores are septate.

### *Cyttariaceae*

The very curious family Cyttariaceae contains only one genus, *Cyttaria*. Six species are known, occurring in New Zealand, Tasmania, and South America; all are parasitic upon species of *Nothofagus*.

*C. Darwinii* occurs very commonly in Tierra del Fuego, where it was collected by Darwin in 1833.

"In the beech forests," he says, "the trees are much diseased; on the rough excrescences grow vast numbers of yellow balls. They are of the colour of the yolk of an egg, and vary in size from that of a bullet to that of a small apple; in shape they are globular, but a little produced towards the point of attachment. They grow both on the branches and stems in groups. When young they contain much fluid and are quite tasteless, but in their older and altered state they form a very essential article of food for the Fuegian. The boys collect them and they are eaten uncooked with fish." He observed that they were smooth when young, "the external surface marked with white spaces as of a membrane covering a cell"; later the whole surface is "honeycombed by regular cells." These are the separate apothecia, considerable numbers of which occur on the same stroma. Bertero, at about the same time, recorded that the Chilian species (*C. Berteroi*) threw

Fig. 87. *Cyttaria Gunnii*, Berk.; *a.* twig of *Nothofagus Cunninghami* with knobs bearing the fungus, × ⅜; *b.* group of stromata; *c.* single stroma cut across; all after Berkeley.

"out of these cavities an impalpable powder when it was touched, exactly as is observed in the *Peziza vesiculosa*."

In all species the stromata seem to grow from a distinct disc (fig. 87), formed from the bark or the bark and wood of the host, and traversed in all directions by the mycelium, which doubtless gives rise to a fresh crop each season. The asci are rather short and cylindrical and contain eight ovoid spores.

In *C. Darwinii* Berkeley observed that the lower part of the stroma was granulated as if beset with a small, black, parasitic *Sphaeria*; Fischer interpreted these structures as spermogonia or pycnidia, and was able to observe them on different parts of the stroma of *C. Hookeri* and *C. Harioti*. He also noted, below the developing apothecia of *C. Darwinii*, certain stout, coiled, branching hyphae, which were at this stage almost without contents. Their appearance suggests that they are ascogenous hyphae, but Fischer made the alternative suggestion that they might be archicarps. In view of the presence of putative spermatia he made some search for trichogynes reaching to the surface of the stroma, but could find none, nor any evidence of a sexual process.

The Cyttariaceae have been compared to those members of the Cenangiaceae in which the apothecia arise from a common stroma, and in which pycnidia or spermogonia are also present.

In many directions they require further investigation.

## CYTTARIACEAE: BIBLIOGRAPHY

1842 BERKELEY, M. J. On an edible Fungus from Tierra del Fuego and an allied Chilian Species. Trans. Linn. Soc. Lond. xix, p. 37.
1847 BERKELEY, M. J. Fungi. Hooker's Flora Antarctica, ii, p. 453.
1848 BERKELEY, M. J. Decades of Fungi. Lond. Journ. Bot. 2nd series, vii, p. 576.
1885 BUCHANAN, J. On *Cyttaria Purdiei*. Trans. New Zealand Inst. xviii, p. 317.
1886 FISCHER, E. Zur Kenntniss der Pilzgattung *Cyttaria*. Bot. Zeit. xlvi, p. 812.

## HELVELLALES

The members of the Helvellales are saprophytes, growing chiefly on the ground, sometimes on decayed wood and branches. Most are large, fleshy and stipitate. The hymenium is spread over the upper surface, and, in the few forms studied, is covered at first by a veil or membrane through which the paraphyses break, much as in the Pezizas, and which may be homologized with their peridium. There are three families:

Ascophore flattened, not stalked — RHIZINACEAE.
Ascophore stalked
  fertile region of head distinct from stalk, ascus opening by a lid — HELVELLACEAE.
  fertile region not always distinct from stalk, ascus opening by a plug — GEOGLOSSACEAE.

### *Rhizinaceae*

The Rhizinaceae are characterized by their unstalked fructification, and include the genera *Rhizina* and *Sphaerosoma*.

*Rhizina* has a flattened, crust-like ascophore, more or less concave below, and attached to the soil by root-like strands of mycelium. The asci cover the upper surface.

*Rhizina inflata* occurs in this country only as a saprophyte, growing on soil, but in both France and Germany it has been found to attack conifers. The disease, known as ring disease, or root fungus, extends from a centre, infecting one plant after another and causing them to lose their needles and die. The mycelium ramifies in the intercellular spaces of the cortex, and within as well as between the cells of the bast, so that the sieve-tubes are completely filled. It forms also masses of pseudoparenchyma between the dead and diseased tissues of the host.

The development of the ascocarp has been studied in *R. undulata* where Fitzpatrick found a long, multicellular archicarp recalling that of some of the Ascobolaceae. He regards the terminal cell or cells as a trichogyne but there is no evidence that normal fertilization ever takes place or that male organs are ever developed. In due course, the central cells become continuous through large pores, and give rise to ascogenous hyphae into which the nuclei pass. Fitzpatrick observed paired nuclei in the oogonial region and in the ascogenous hyphae, and infers that nuclear association occurs in the archicarp, but that there is only one fusion, that in the ascus.

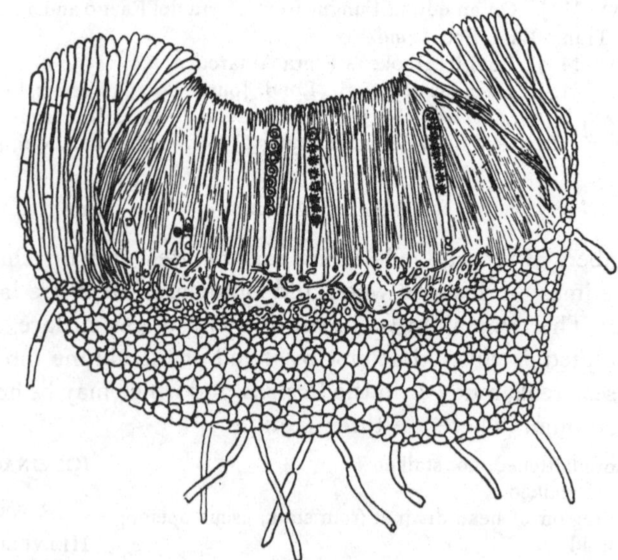

Fig. 88. *Sphaerosoma Janczewskianum* Roup.; apothecium showing oogonial cell, × 70; after Rouppert.

In *Sphaerosoma* the ascophore is more or less sunk in the substratum, and is attached by rooting hyphae which are sometimes grouped on a short pedicel. It is concave when young, but later forms an irregularly globose mass over the upper surface of which the hymenium is spread (fig. 89). It resembles, in fact, a *Peziza* which becomes very much reflexed at maturity.

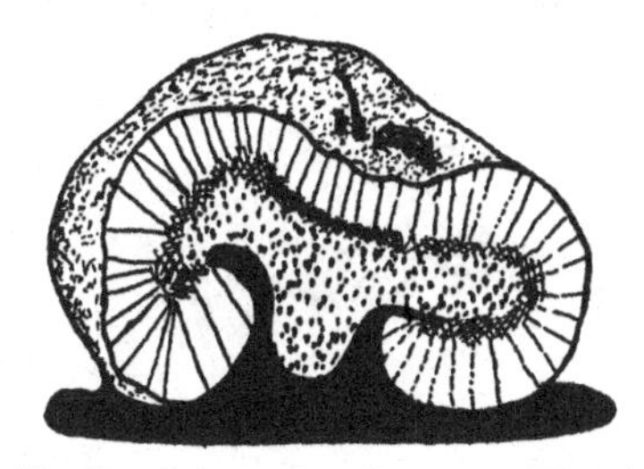

Fig. 89. *Sphaerosoma fuscescens* (Klotz.) Roup.; apothecium, × 6; after Rouppert.

In *Sph. Janczewskianum* (fig. 88), a large oogonial cell has been recognized from which the ascogenous hyphae originate, but no details of its development are known. This genus has been variously placed in the Tuberaceae and Pezizaceae as well as in its present position. It shows resemblances to some of the former in its habitat under fallen leaves, and to the latter group in many points of general structure.

## RHIZINACEAE: BIBLIOGRAPHY

1909 ROUPPERT, C. Revision du genre Sphaerosoma. Bull. Acad. Sci. de Cracovie, p. 75.
1918 FITZPATRICK, H. M. Sexuality in *Rhizina undulata* Fries. Bot. Gaz. lxv, p. 201.

## *Helvellaceae*

The Helvellaceae are represented by five genera, *Helvella* (fig. 90 *a*), *Morchella* (fig. 90 *b*), *Verpa*, *Gyromitra*, and *Cidaris*; of these the first four are British. In all a definite fertile head is distinguished from the sterile stalk, and over the more or less convoluted surface of the head the hymenium extends.

Development has been studied only in species of *Helvella* where the fruit arises as a tuft of branching, septate hyphae, and no archicarp has been observed.

In *H. elastica*, young ascophores, about 0·5 mm. in diameter, show no signs of fertile hyphae. A membrane of interwoven filaments encloses the whole fruit body, and below this a palisade of club-shaped hyphae is differentiated. As growth proceeds the membrane becomes broken, and the palisade increases in regularity, forming the boundary of the fructification except where, at the apex, the paraphyses are growing up. Later, as these increase in number, the ascogenous hyphae appear among them, and numerous asci are formed.

In *H. crispa* the later stages of development are very similar to those in *H. elastica*. Here nuclear fusions have been observed in the young ascogenous hyphae, replacing, as in *Humaria rutilans*, the obsolete sexual fusions, and preceding the fusions in the asci. Carruthers has studied the

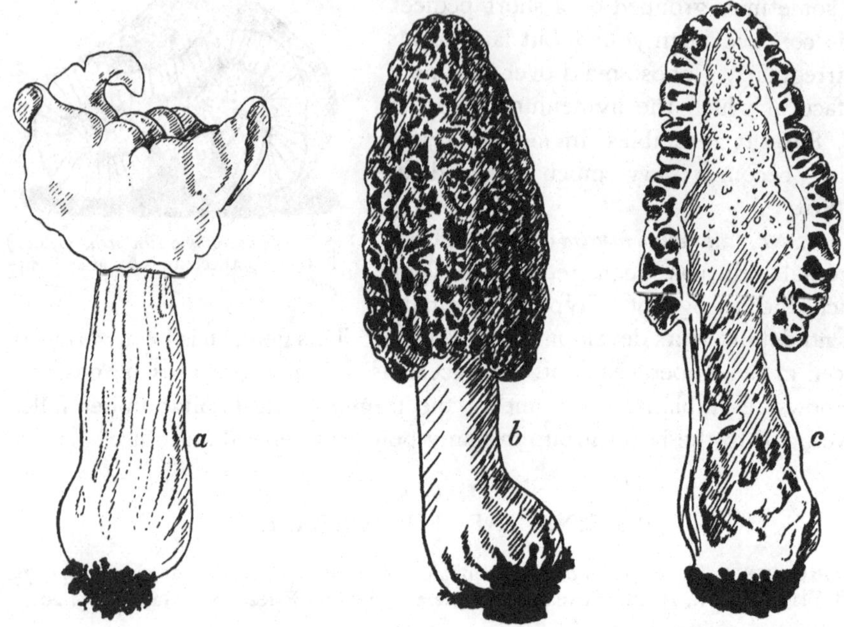


Fig. 90. *a. Helvella crispa* (Scop.) Fr.; *b.* and *c. Morchella vulgaris* Pers.; after Boudier.

nuclear divisions, and finds two chromosomes in the vegetative and four in the fertile hyphae. Four again appear in the first and second (meiotic) divisions in the ascus, after the second fusion has taken place, and two are recorded in the telophase of the third division, and in the mitosis in the spore. The ripe spore normally contains eight nuclei.

In both species, after an ascus has arisen from the penultimate cell of a hypha, the terminal cell may grow on, giving rise to others, and may fuse before doing so with the third cell of the hypha, which is the stalk-cell of the previously formed ascus.

In *Morchella esculenta* the nuclear divisions of the ascus have been studied by Maire. After observing eight chromatin bodies in the prophase of the first division in the ascus, he found four in the prophase and anaphase of the third, and in the divisions of the spore nuclei; this corresponds closely with Carruthers' results in *Helvella crispa.*

## HELVELLACEAE: BIBLIOGRAPHY

1905 MAIRE, R. Recherches cytologiques sur quelques Ascomycètes. Ann. Myc. iii, p. 123.

1910 MCCUBBIN, W. A. Development of the Helvellinaceae. I. *Helvella elastica.* Bot. Gaz. xlix, p. 195.

1911 CARRUTHERS, D. Contributions to the Cytology of *Helvella crispa.* Ann. Bot. xxv, p. 243.

*Geoglossaceae*

The Geoglossaceae grow usually in damp or moist situations such as low, wet woods and shady slopes. They occur on soil or on dead branches or leaves, and two species of *Mitrula* are parasitic on living moss. The family includes some eight genera of which five are British.

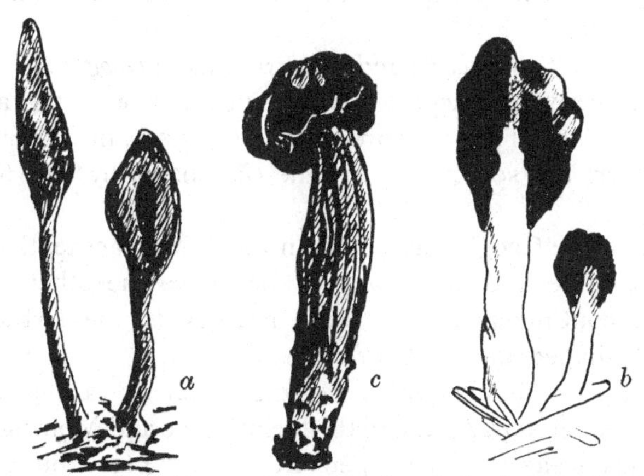


Fig. 91. *a. Geoglossum hirsutum* Pers., nat. size; *b. Spathularia clavata* Sacc., nat. size; *c. Leotia lubrica* Pers., form *stipitata*, × ⅜; after Massee.

The ascophore is erect and stipitate with the fertile portion terminal, and either club-shaped (fig. 91 *a, b*), laterally compressed, or forming a cup or a pileus (fig. 91 *c*.). In some of the simpler forms, as in *Geoglossum hirsutum*, there is no clear line of demarcation between the fertile and sterile regions. The ascus contains eight spores and opens by the ejection of a plug.

The young ascocarp consists of a dense tangle of vegetative filaments; in the early stages a more or less conspicuous veil has been identified in several genera (though not as yet in *Geoglossum*). It is composed, as in the Helvellaceae, of interwoven hyphae, derived from and continuous with the outer layer of the fruit body. There are indications that it opens at first by a pore at the apex, but it soon breaks up into scales and disappears.

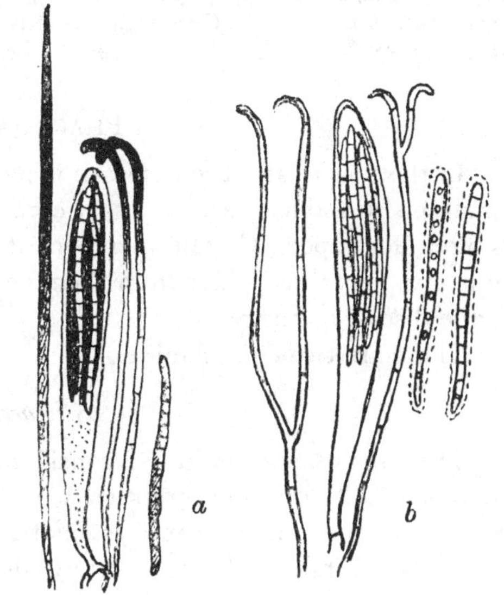


Fig. 92. *a. Geoglossum hirsutum* Pers., × 230; *b. Spathularia clavata* Sacc., × 400; after Massee.

In *Leotia lubrica* a large branching cell, presumably an oogonium, occurs at the base of the very young ascocarp and appears to give rise to the ascogenous hyphae.

As far as the characters of the mature fruit are concerned, two lines of development can be traced, both starting from *Geoglossum* and passing, the one through *Spathularia* to *Vibrissea*, the other through *Mitrula* and *Leotia* to the Helvellaceae.

In the species of *Spathularia* and *Vibrissea*, as in *Geoglossum*, the spores are very long, narrow and septate, lying side by side in the ascus. *Geoglossum* is distinguished by its coloured spores (fig. 92 *a*), the other two genera, in both of which the spores are hyaline (fig. 92 *b*), by the form of the fructification.

In the rest of the Geoglossaceae, as in the Helvellaceae, the spores are elliptical and hyaline, and are arranged one above the other in the ascus. They may be continuous or septate. In *Mitrula* the fertile region is irregularly club-shaped, and in *Leotia* pileate.

A relationship to the Pezizales suggests itself at various points, and perhaps especially through *Leotia*, to the Helotiaceae and Mollisiaceae where, as in the Geoglossaceae, the ascus opens by a slit or pore from which a plug of wall substance is ejected, not as in the majority of the Helvellales and Pezizales by a definite lid.

## GEOGLOSSACEAE: BIBLIOGRAPHY

1897 MASSEE, C. A Monograph of the Geoglossaceae. Ann. Bot. xi, p. 225.
1908 DURAND, E. J. The Geoglossaceae of North America. Ann. Myc. vi, p. 387.
1910 BROWN, W. H. The Development of the Ascocarp of *Leotia*. Bot. Gaz. l, p. 443.

## PHACIDIALES

In the Phacidiales the ascocarp is immersed in the matrix. It is usually small in size and leathery, waxy, or coriaceous in consistency; an epithecium is often developed. Certain members of the group resemble the Hysteriales in many points and differ from them chiefly in the greater exposure of the fertile disc at maturity.

There are two chief families.

### *Stictaceae*

The Stictaceae constitute a considerable group of small forms, occurring saprophytically on wood or other plant remains. Their development and minute anatomy, apart from systematic characters, is practically uninvestigated. They have a fleshy or waxy disc, pale and clear coloured, usually white, yellow, or tinged with pink. The sheath is not always developed, when present it is thin and white and is mealy owing to the presence of particles of calcium oxalate; when the fruit opens it forms a white border around the hymenium. The pale colour, and the ragged or toothed dehiscence of the sheath are very characteristic.

### *Phacidiaceae*

The Phacidiaceae are distinguished by their black, thick-walled apothecia, usually scattered, sometimes, as in *Rhytisma*, grouped on a black stroma. Where the fertile disc is circular the sheath splits in a stellate manner, but where it is elongated, dehiscence takes place by means of a slit running along its entire length. The species occur chiefly on dead herbaceous stems or leaves, but a few are parasitic.

*Rhytisma Acerinum* (fig. 93) infects the leaves of various species of *Acer* (maple and sycamore). The mycelium ramifies in the living tissues of the

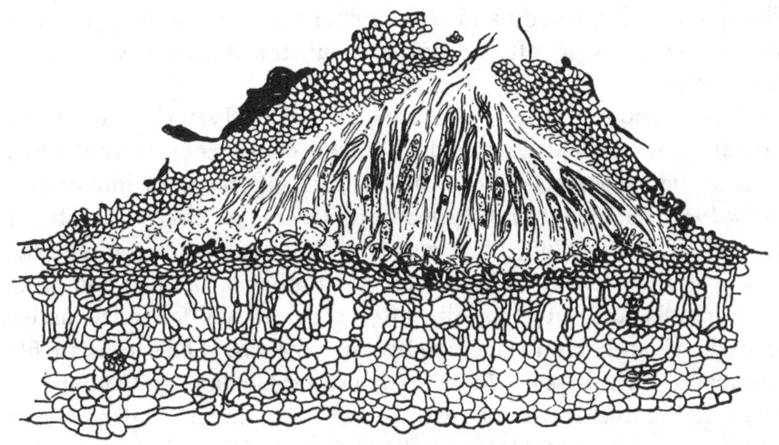

Fig. 93. *Rhytisma Acerinum* (Pers.) Fr.; apothecium, × 160.

host and causes yellow spots on the leaves about three weeks after infection. Some five weeks later pycnidia develop under the cuticle and produce small unicellular conidia. The epidermis and underlying tissues of the host become filled with hyphae and a dense, black sclerotium is completed. In this state the leaf falls and next spring the sclerotia thicken and become wrinkled; finally they burst by elongated fissures and expose the discs of the apothecia. The ascospores are filiform and septate; they are ejected with some force and reach the living leaves to which they are probably carried by the wind.

## Hysteriales

The Hysteriales are characterized by the black, elongated ascocarp, dehiscing by a longitudinal slit, so narrow that the disc is almost permanently concealed.

The species are all minute; in some the disc is narrowly elliptical, in some it branches in a stellate manner, in others the ascocarp is raised and laterally compressed so as to resemble a miniature mussel or oyster shell standing on its hinge and with the opening uppermost. In this case, or when the ascocarp is superficial, it is rigid and carbonaceous in consistency, when developed beneath the epidermis of the host it is membranous.

The ascospores are coloured or hyaline and are frequently septate; they may be very long and narrow and may be surrounded by a gelatinous membrane.

In a few cases pycnidia are known, producing oblong, unicellular, hyaline conidia.

The majority of species are saprophytic on old wood, bark, or dry leaves. The mycelium is intercellular, and is sometimes parasitic on living plants though the apothecia reach maturity only on parts that have been killed.

The details of cytology and development are not known, nor do these minute species, growing often on a hard substratum, seem very promising objects of study.

The subdivisions of the Hysteriales, of which Hysteriaceae and Hypodermataceae are the chief, depend upon the consistency of the sheath, on the form of the ascocarp, and on whether it is superficial or immersed.

*Lophodermium Pinastri* (Hypodermataceae) produces pine-blight or needle-cast in the seedling of *Pinus sylvestris* and other conifers, causing them to drop their leaves. The mycelium ramifies in the leaf and gives rise first to pycnidia and later, usually after the leaf has fallen, to ascocarps. These are black and oblong, opening by a narrow slit. The spores are filiform and continuous. The disease does very considerable damage to young plants, often causing death. It attacks mature trees also and, though these are not themselves seriously injured, they act as centres of infection, particularly in the neighbourhood of seedbeds and nurseries.

In the form of their fructification the Hysteriales are intermediate between the Discomycetes on the one hand, and the Pyrenomycetes on the other, and have been variously included under either of these headings. Their black, coriaceous ascocarps, opening by a narrow slit, differ from those of certain Phacidiales chiefly in the less exposure of the disc.

They approach the Sphaeriales in the frequent occurrence of coloured, septate spores, as well as in the consistency and often in the form of the ascocarp, which is distinguished from a true perithecium chiefly by its elongated opening, and by the absence of periphyses[1]. Possibly a study of their minute anatomy may lead to more definite knowledge of their relationships.

[1] For definition, see p. 140.

## TUBERALES

The Tuberales are typically subterranean though some species are only imperfectly buried, or grow among decaying leaves. When mature the fruits emit a powerful odour by which rodents are apprised of their whereabouts. The ascocarp is eaten and the spores dispersed after passing through the alimentary canal of the animal.

The ascocarp is more or less globose, sometimes completely closed, sometimes with a small opening. The hymenium may form a smooth lining to the fruit or may be thrown into elaborate folds so that the fertile region is divided into chambers. The asci contain one to eight spores, but, as far as is known, eight nuclei are always produced. The epispore is often elaborately ornamented at maturity.

Early investigators classed the Tuberaceae with the hypogeal Gasteromycetes, and a consequence of this survives in the use of the term **gleba** to describe the contents of the ascocarp, including both vegetative hyphae and hymenium.

The Tuberales include a single family, the Tuberaceae; their relationship is probably to the Pezizaceae and Rhizinaceae. One or more series can be traced between these families and the truffles, the principal modifications being in the direction of adaptation to subterranean conditions by increased protection of the hymenium. This appears to have been achieved either by retaining the closed form of the young pezizaceous apothecium (*Genea*, *Pachyphloeus*) or by invagination of the fertile layer (*Tuber*) over a widely exposed surface such as is found in *Rhizina* or *Sphaerosoma*. In either case room has been made for additional asci by throwing the hymenium into elaborate folds. Massee, however, regards the globose asci and dark-coloured sculptured spores of *Tuber* as primitive, and derives from it *Genea*, and thence the Pezizales.

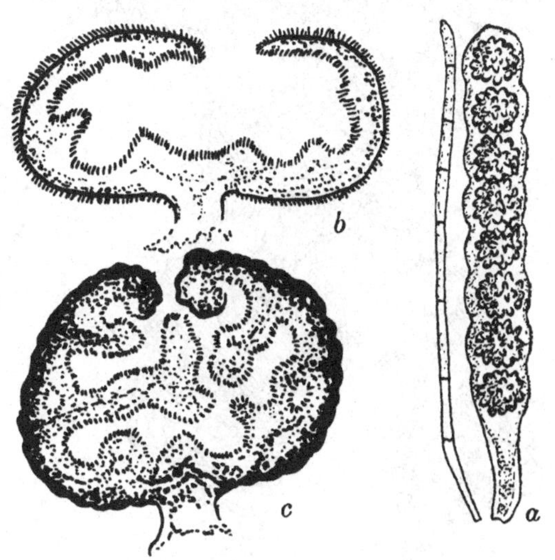


Fig. 94. *a*. *Genea Klotzschii* B. and Br.; ascus and paraphysis; *b*. *Genea hispidula* Vitt.; apothecium; *c*. *Genea sphaerica* Vitt.; apothecium; after Massee.

### *Tuberaceae*

In *Hydnocystis* and *Genea* the ascocarp is fleshy or warted; it has a single aperture often more or less closed by projecting hyphae. Internally the hymenium may form a smooth lining, or, in *Genea*, is more often divided into chambers, all of which communicate with the apical opening. The asci are cylindrical and contain eight uniseriate spores (fig. 94 *a*). The simplest species in fact resemble a nearly closed *Peziza* (fig. 94 *b*, *c*).

In *Stephensia* and *Pachyphloeus* the hymenium is more elaborately convoluted; the asci in *Pachyphloeus* are stouter, and the spores irregularly biseriate.

In *Balsamia* (figs. 95, 96) the asci are broadly oblong or subglobose; the mature ascocarp is completely closed and surrounded by a pseudoparenchymatous sheath. The youngest ascocarps of *B. platyspora* which

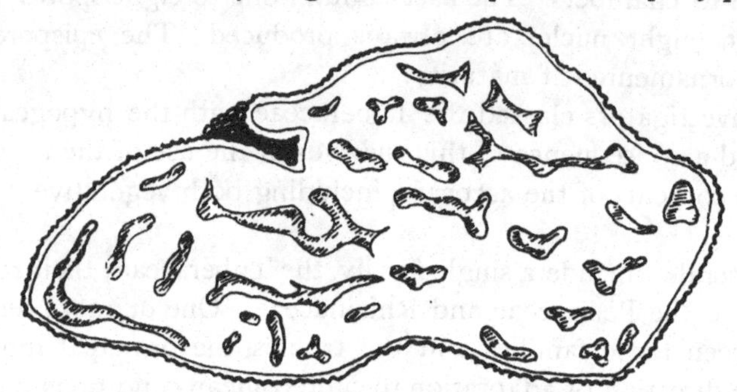

Fig. 95. *Balsamia vulgaris* Vitt.; after Tulasne.

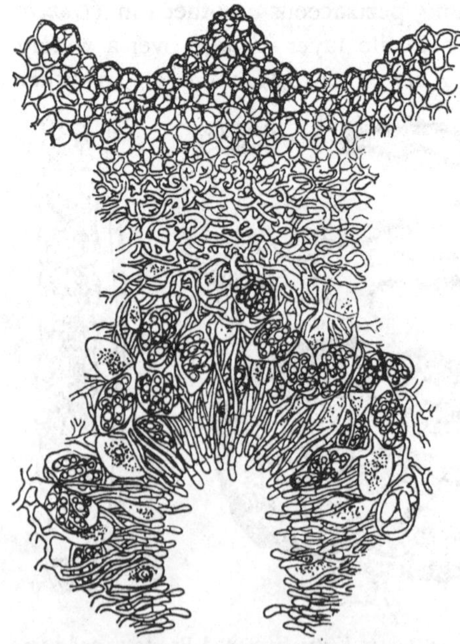

Fig. 96. *Balsamia vulgaris* Vitt.; section through hymenium; after Tulasne.

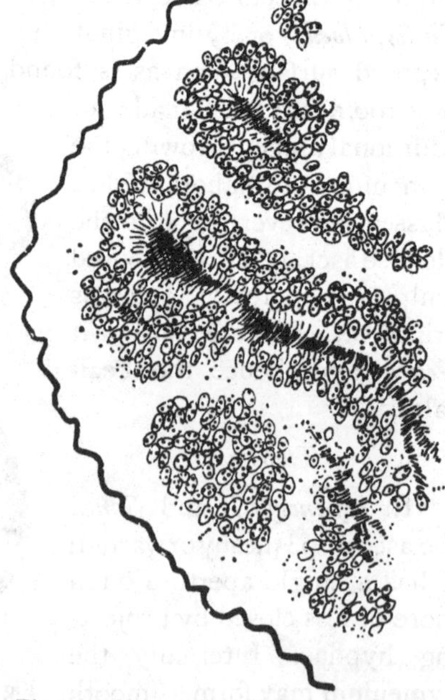

Fig. 97. *Tuber rufum* Pico; general view of fertile region; after Tulasne.

Bucholtz was able to examine, showed a system of internal chambers lined by the hymenium and communicating at one or more points with the exterior. As development proceeds these cavities increase in size and the hymenium becomes further convoluted, so that additional chambers are formed.

In *Tuber* the ascocarp is irregularly globose, fleshy or sometimes almost woody; internally the walls which divide the gleba are extensively branched, and the free space between them is diminished, so that the layers of the hymenium are brought close together and constitute the fertile "veins." Other "veins," white and sterile, run between the hymenial layers and serve as air chambers (fig. 97). The asci are often globose, and the spores usually four in number, but the number varies, and is sometimes reduced to two or one (fig. 98).

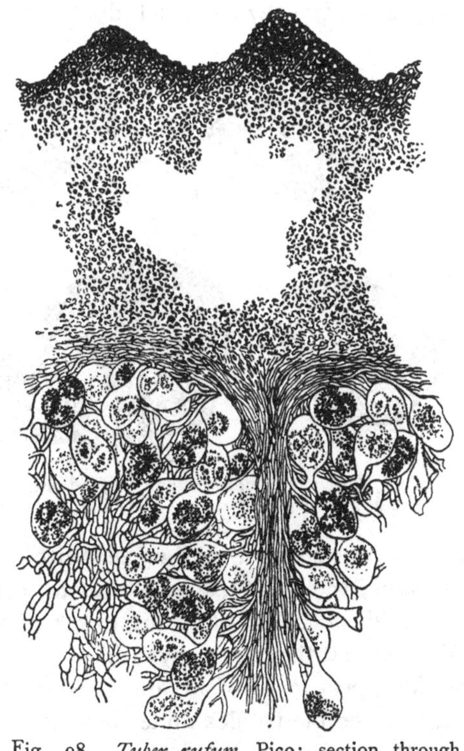

Fig. 98. *Tuber rufum* Pico; section through hymenium; after Tulasne.

The development of the fruit has been studied by Bucholtz in *Tuber puberulum* (fig. 99). The very young ascocarp consists of a mass of hyphae, the outer rather more loosely interwoven than the inner. Around the lower part a dense basal sheath is differentiated. Soon the first signs of the fertile veins appear as invaginations of the upper surface, and internally the loose tissue of the sterile veins becomes recognizable.

Owing to the rapid growth of the upper portion of the young fruit, the basal sheath is bent backwards, while at various points along the fertile veins the first signs of asci appear. Later the peripheral tissues become thickened, together with the remains of the basal sheath, and form the peridium. This ultimately closes over the points where the fertile veins are in communication with the exterior. Thus the young fruit is open at first, the hymenium becomes internal by invagination and the peridium which covers the mature ascocarp is a secondary formation.

The development of the fructification in *Choiromyces maeandriformis* approaches that of *T. puberulum*, but the basal sheath and peridium are less conspicuous.

The ascocarps of many species of *Tuber* are edible, the most esteemed being *T. melanosporum* which does not occur in Britain. They grow chiefly

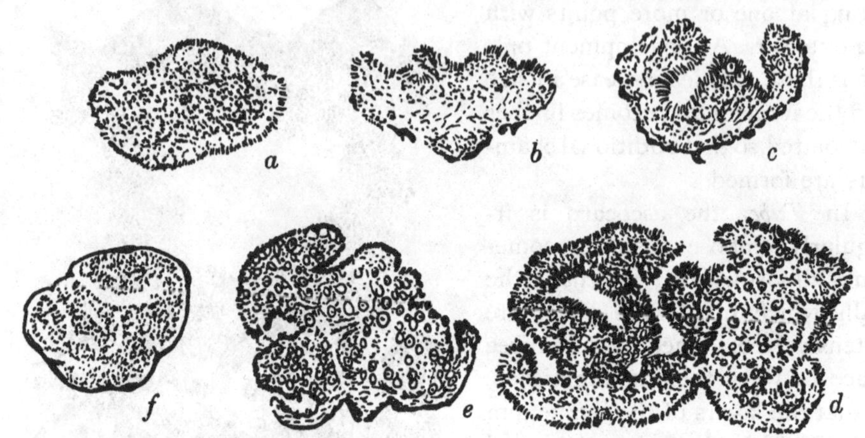


Fig. 99. *Tuber puberulum* (B. and Br.) Ed. Fisch.; *a.—e.* development of ascocarp; *a.* × 52; *b.* and *c.* × 28; *d.* and *e.* × 21; *f.* section through mature ascocarp, × 6; all after Bucholtz.

in soils consisting of sand mixed with clay and containing iron, or in mixed alluvium; the soil must be porous to secure sufficient aeration.

Truffles occur in chestnut, oak, and especially beech woods and there is evidence that they form mycorhiza with the roots of these trees. The relation would appear to be of advantage to the fungus since the success of the cultivation of edible truffles under oaks in France depends on keeping the roots near the surface.

## TUBERACEAE: BIBLIOGRAPHY

1903 BUCHOLTZ, F. Zur Morphologie und Systematik der Fungi hypogaei. Ann. Myc. i, p. 152.

1905 FAULL, J. H. Development of Ascus and Spore Formation in Ascomycetes. Proc. Boston Soc. Nat. Hist. xxxii, p. 77.

1906 BOULANGER, E. Notes sur la Truffe. Soc. Myc. xx–xxii, pp. 77 etc.

1908 BUCHOLTZ, F. Zur Entwickelung der Choiromyces Fruchtkörper. Ann. Myc. vi, p. 539.

1909 MASSEE, G. The Structure and Affinities of British Tuberaceae. Ann. Bot. xxiii, p. 243.

1910 BUCHOLTZ, F. Zur Entwickelungsgeschichte des Balsamiaceen-Fruchtkörpers nebst Bemerkungen zur Verwandtschaft der Tuberineen. Ann. Myc. viii, p. 121.

# CHAPTER V

## PYRENOMYCETES

THE Pyrenomycetes include some 10,000 species; they are characterized by the fact that their ascocarp or perithecium is a more or less flask-shaped organ opening by a narrow pore, the ostiole, and containing a hymenium spread in a regular manner over the floor and lower part of the sides (fig. 100). It thus differs from the perithecium of the higher Plectascales where the asci are irregularly scattered, and from that of the Erysiphales where, except in the flattened perithecium of the Microthyriaceae, an ostiole is not developed. By some authors the term Pyrenomycetes is used to include all these groups and even certain other forms, such as the Tuberales. A study of the development of the truffles, however, has made clear their affinity with the Pezizales; the mildews constitute a well defined and isolated group, distinguished, so far as they are known, by the form of their sexual organs; and the higher Plectascales differ from the present series and resemble the simpler forms with which they have here been classified in the important character of the arrangement of their asci.

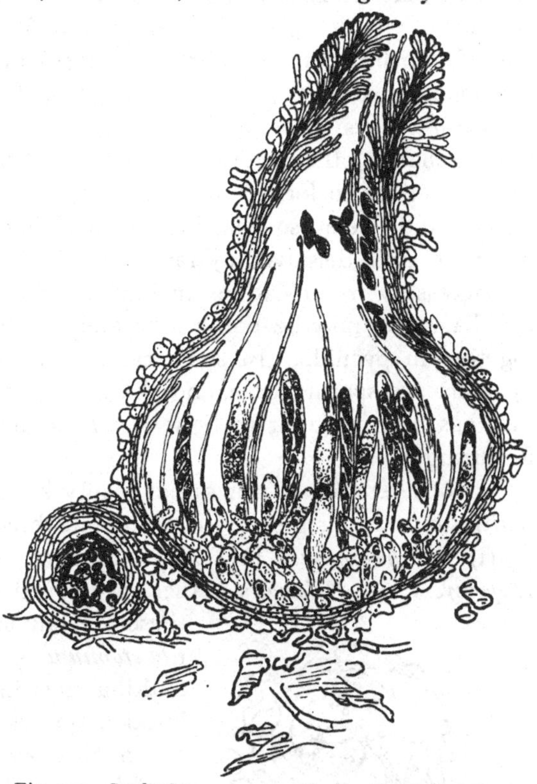

Fig. 100. *Sordaria* sp.; ascocarp in longitudinal section showing asci, paraphyses and periphyses, × 400.

There remain four groups, the Hypocreales, the Dothideales, the Sphaeriales, and the Laboulbeniales:

The last are true Pyrenomycetes in the sense that they possess regularly arranged asci and a perithecium opening by an ostiole, but they are distinguished by so many special characters that, though included under this heading, they can best be dealt with apart.

The Hypocreales, Dothideales and Sphaeriales, have in common more or less pyriform or flask-shaped perithecia; these are sometimes isolated and free, sometimes sunk in the tissue of the host, and sometimes embedded in a **stroma** or cushion of fungal tissue. The perithecium is lined by delicate filaments, some of which, the **periphyses**, grow along and partially close the neck, and may protrude through the ostiole, while others (paraphyses) are mingled with the asci in the venter of the fruit. The neck of the perithecium varies very much in length, and is often markedly phototropic, the ostiole being directed towards the light, and thus incidentally towards a clear space so that, when the spores are shed, as wide a distribution as possible is ensured. So definite is this reaction in, for example, species of *Sordaria*, that if the direction of light be changed every four or five days during development, a series of corresponding bends in the neck are produced. In *Sordaria* and its allies the asci elongate, reaching up to the ostiole and in turn discharging their spores; in species of *Spumatoria* and *Chaetomium* the asci deliquesce to form a mucilaginous mass which readily absorbs water and expands, being squeezed up the neck and exuded at the ostiole where it persists until dissolved by rain or dew.

Accessory fructifications include chlamydospores and various types of conidia which may be borne separately on free conidiophores, or grouped together in pycnidia. In some cases there is evidence that the so-called pycnidia are spermogonia, and the spores they produce spermatia, but no case has been brought to light in which these still fulfil their function as fertilizing agents.

A consideration of our rather scanty knowledge of the initiation of the perithecium in this group brings to light three main types of development.

(i) In *Chaetomium*, in *Sordaria* (fig. 101) and its allies and in species of *Hypomyces* and *Melanospora* there is a coiled archicarp of four or five cells; these are uninucleate in *Hypomyces lateritus*, *Chaetomium spirale* and *Podospora hirsuta*, multinucleate in *Sordaria* and *Hypocopra* and in other species of *Chaetomium*.

In *Sordaria macrospora* the archicarp is straight instead of coiled and in *S. fimiseda* a swollen terminal cell has been reported. A pair of initial hyphae has been described in *Rosellina quercina*, but in no case has a sufficiently detailed study been made either to reveal nuclear fusions in the archicarp or to justify the inference that they do not occur. Under these circumstances it is possible to judge of the function of these initial filaments

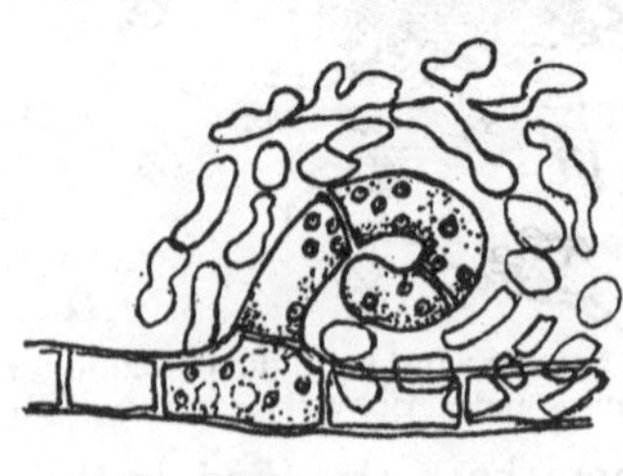

Fig. 101. *Sordaria fimicola* Rob.; archicarps; after Dangeard.

only on the somewhat incomplete evidence that they give rise to ascogenous hyphae and on the basis of their resemblance to the sexual branches of other Ascomycetes. There is a very similar coiled hypha in certain species of *Eurotium* which is certainly a functional archicarp. Comparison may also be made with the female branch of *Ascodesmis nigricans* and with that of the Erysiphaceae.

(ii) The second type of pyrenomycetous initial organ (fig. 102) may

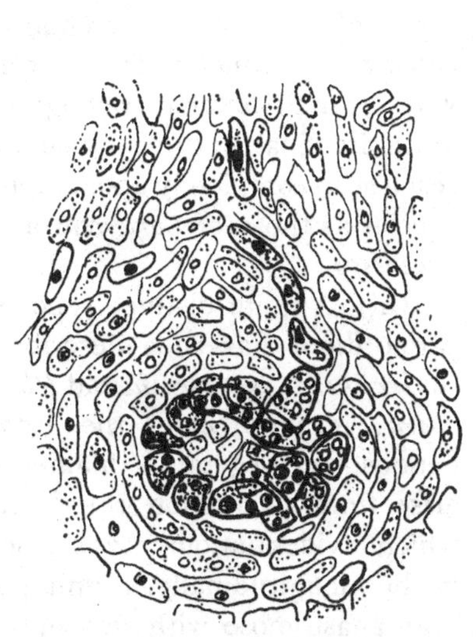

Fig. 102. *Polystigma rubrum* DC.; mature archicarp, × 800; after Blackman and Welsford.

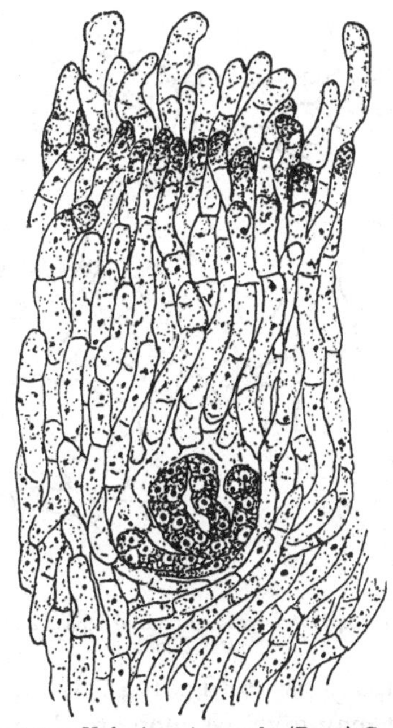

Fig. 103. *Xylaria polymorpha* (Pers.) Grev.; archicarp embedded in stroma, × 1000.

readily be derived from the first. It occurs in forms where the perithecium is immersed either in the substratum or in a stroma, and its essential character is the prolongation of the tip of the archicarp to form a trichogyne-like organ. The appearance of this structure is associated with the development of spermatia in spermogonia. Archicarps of the type in question are found in *Polystigma* among the Hypocreales and in *Gnomonia*, *Poronia* and *Mycosphaerella* among the Sphaeriales. In all these genera, however, the trichogyne appears to be merely vestigial; in *Polystigma* it never reaches the exterior of the host-leaf, in *Gnomonia* its connection with the coiled oogonial region is doubtful and in *Mycosphaerella* and *Poronia* it degenerates early. In *Polystigma* the ascogenous hyphae arise from vegetative cells and not from the archicarp and it is at least possible that the same is the case in the other genera named. A comparison is obvious between the archicarps of these forms and those of several Lichens and of such Discomycetes as

*Lachnea cretea* and the *Ascoboli*, where the coiled and septate archicarp is often still functional.

A very common initial organ in forms with embedded perithecia is the short filament of cells sometimes known as Woronin's hypha (fig. 103). The cells are large and contain well-marked nuclei and lie in a nest of small-celled vegetative mycelium. Woronin's hypha has been found among the Hypocreales in *Nectria* and among the Sphaeriales in *Xylaria* and *Hypoxylon*; it remains to be shown whether it still functions. It may have originated from the simple archicarps of the Lower Pyrenomycetes or by reduction from forms with a multicellular trichogyne. With its final disappearance we reach such completely apogamous species as those of *Cordyceps* and *Claviceps*.

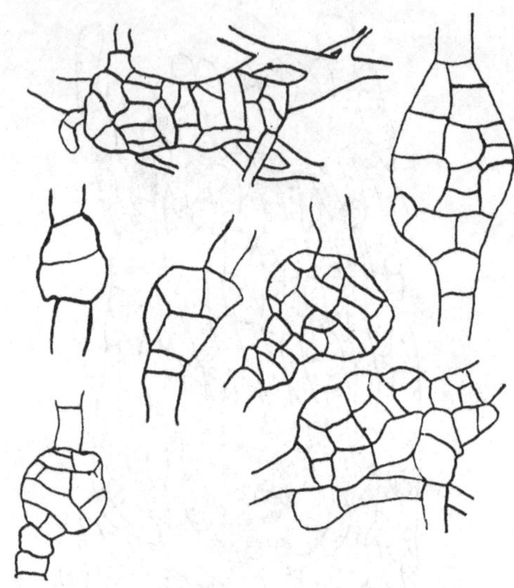

Fig. 104. *Strickeria* sp.; initial cells of ascocarps; after Nichols.

(iii) A quite distinct type of primordium has been described in *Strickeria*, *Sporormia* and *Pleospora*; in these cases the ascogenous and vegetative filaments arise from a common initial cell which divides not only transversely, but longitudinally, forming a compact tissue (fig. 104). Other hyphae may anastomose with this mass, or it may give rise alone to the whole fructification. Possibly some suggestion of its origin may be found in the peculiar, but apparently normally fertilized oogonium of *Leptosphaeria*.

The Pyrenomycetes do not appear to have given rise to any higher forms, and have themselves a greater vegetative development than any other Ascomycetes.

They may be subdivided as follows:

| | |
|---|---|
| Wall of perithecium differentiated from stroma; | |
| perithecium wall and stroma, if present, soft in texture, either colourless or light coloured | HYPOCREALES. |
| perithecium wall and stroma, if present, firm, leathery or brittle, dark in colour | SPHAERIALES. |
| Perithecium always sunk in a stroma from the tissue of which its wall is not differentiated; colour of stroma black or dark brown | DOTHIDEALES. |
| Minute, external parasites on insects, perithecium borne on a receptacle which also bears appendages; spores two-celled | LABOULBENIALES |

## HYPOCREALES

The Hypocreales are readily distinguished by the clear colour and more or less fleshy consistency of the perithecium or stroma. In the majority of Pyrenomycetes the colour is black or dark brown, but here bright red, yellow, blue, and various paler shades are found, and it is only quite occasionally that so dark a tint as brown or dirty violet appears.

The asci contain usually eight, sometimes four, and sometimes many spores. The spores are in most cases hyaline, but are dark-coloured in *Melanospora* and its allies; they are elliptical or filiform, and may be one or more celled.

In a number of species conidia as well as ascospores are produced.

The group includes both saprophytic and parasitic forms.

There are some sixty genera of Hypocreales, and the group is subdivided primarily according to the development of the stroma. In the simplest forms the stroma is absent, and the separate perithecia may or may not be partly sunk in the substratum, in others a filamentous or a fleshy stroma appears, and the perithecia are more or less embedded. In the highest members the perithecia originate deep in the stroma, and remain immersed in it throughout their development.

Upon these characters the subdivision of the group is based:

Stroma absent, or, when present, with perithecia entirely superficial ... NECTRIACEAE.
Stroma forming a conspicuous matrix in which the perithecia are partially or entirely immersed ... HYPOCREACEAE.

### *Nectriaceae*

The species of the genus *Hypomyces* are for the most part parasitic upon the pilei of various Hymenomycetes. *H. aurantius* occurs on old *Polypori* and on species of *Stereum*.

Here the free perithecia are roughly oval in form, orange yellow in colour, and seated on a delicate filamentous stroma. The perithecium wall consists of an outer coat of narrow, closely woven hyphae, and an inner layer of larger, thinner-walled cells with scanty contents. The cavity becomes filled with paraphyses and developing asci, and is prolonged into the neck lined with short periphyses. The spores are two-celled and the wall at each end is usually prolonged into a point.

Development has been studied by Moreau in *Hypomyces lateritus*, a form parasitic on species of *Lacterius*, and placed by Maire in the genus *Peckiella* by reason of its unicellular spores. The cells of the vegetative mycelium are uninucleate, and the archicarp appears among them as a coil of uninucleate cells; there is no sign of an antheridium. Nuclear division without wall-formation takes place in the archicarp so that each cell contains two or occasionally three nuclei. At a later stage, after abundant branching, the young perithecium contains a number of binucleate cells; from these the asci arise, the hypha bending over and cutting off a binucleate subterminal cell in the usual way. No fusion but that in the ascus was observed by Moreau. The chief interest of this life history lies in the origin of the binucleate condition, as in some Basidiomycetes, by nuclear division.

In *Melanospora* the stroma may be absent, but when present is characteristically fleshy; the perithecium neck is elongated; the species occur on the fructifications of the Pyrenomycetes, on those of Pezizaceae and Tuberaceae, on various plant remains and in one or two cases on living plants; thus *M. damnosa* may be a serious disease on wheat and rye.

The development of *M. parasitica* was studied by Kihlman; this species is a parasite on certain fungi parasitic on insects, including *Cordyceps militaris*, which is itself a member of the Hypocreaceae.

The first sign of the development of perithecia is the yellow coloration of the mycelium, which has hitherto been white.

The archicarp is a stout, twisted, multicellular hypha forming two or more coils and ending in a somewhat pointed cell; its growth is renewed after the development of the sheath has begun and it divides into some fifteen cells; one of these, which may be termed the ascogenous cell, divides in three directions, forming a true tissue from which the asci arise.

*Melanospora Zobelii*[1] is parasitic on various fungi and especially on the disc of certain Pezizaceae; Nichols found that the spores germinate to give rise to a mycelium in the cells of which the nuclei are arranged more or less in pairs[2]. The archicarp is a coiled or curved branch which becomes septate; near it a more slender antheridial hypha may develop, and, in some cases, may fuse with the female organ.

Vegetative hyphae give rise to a sheath of the usual type with an outer layer, two or three cells thick, of thick-walled, isodiametric cells, and an inner layer of laterally compressed tubular cells; within this is a loose spongy parenchyma of cells rich in contents from which the asci arise. A connection between the archicarp and the asci has not been traced, but it seems probable that the whole central tissue of the perithecium may be derived from the divisions or branches of the archicarp perhaps, as in *M. parasitica*, by means of an ascogenous cell.

A further study of the development of the perithecium and especially of the origin of the asci in this genus is much needed. The facts, especially in *M. parasitica*, suggest that the divisions of the archicarp after the develop-

[1] *Melanospora Zobelii* (Corda) Fuckel = *Ceratostoma brevirostre* Fuckel.

[2] Presumably owing to rapid division; cf. p. 47, *ante*.

ment of the sheath has begun, may correspond to the septation of the fertilized oogonium in other forms. Further, the origin of the asci from a single cell points to the Erysiphales and Laboulbeniales, and in view of the longitudinal divisions, perhaps especially to the latter.

In *Nectria* the usually red or yellow perithecia are produced in groups on stromata of the same colour; the asci contain eight ascospores which are two-celled, and often produce conidia by budding while still in the ascus. The genus is large, including some 250 species among which *N. cinnabarina*, the commonest in this country, is of very frequent occurrence on the living and dead branches of deciduous trees. The mycelium from the germinating spores is unable to penetrate the bark of the host, and infection takes place only through open wounds. Once established, however, the mycelium spreads rapidly especially in the xylem. The cambium and other tissues are not attacked but die as a result of the destruction of the wood, so that as development proceeds branch after branch is killed. Meanwhile the stromata appear (fig. 105); in the conidial stage they are bright pink and occur at all seasons

Fig. 105. *Nectria cinnabarina* (Tde.) Fr. on a fallen twig; *a.* conidial stroma; *b.* young perithecia; ×6; E. J. Welsford del.

on the dead and living branches; perithecia are produced only in the autumn and winter and only after the tissues have been killed; they are deep red in colour and are partly immersed in the deep red stromata. When a perithecium is about to be formed a coil of hyphae larger than the ordinary filaments of the stroma appears a little below the surface, and probably represents the remains of whatever sexual apparatus originally gave rise to the ascogenous hyphae.

*Nectria cinnabarina* is thus one of the rather numerous fungi which produce conidia during their parasitic phase, and ascospores only when the death of the host has rendered them saprophytic. In view of the life-history of this species it is obvious that there are two methods of checking the damage which it does; the burning of infected branches on which the development of the spores takes place, and the painting over of open wounds through which alone the entrance of the mycelium is effected.

## NECTRIACEAE: BIBLIOGRAPHY

1882 MAYR, H. Ueber den Parasitismus von *Nectria cinnabarina*. Untersuchungen aus der forstbot. Institut zu München, iii, p. 1.
1885 KIHLMAN, O. Zur Entwickelungsgeschichte der Ascomyceten *Melanospora parasitica*. Acta Soc. Sci. Fennicae, xiv, p. 313.
1896 NICHOLS, M. A. The Morphology and Development of certain Pyrenomycetous Fungi. Bot. Gaz. xxii, p. 301.
1909 MASSEE, G. On a New Genus of Ascomycetes (*Gibsonia*). Ann. Bot. xxiii, p. 335.
1909 SEAVER, F. J. Notes on North American Hypocreales. Mycologia, i, p. 41.
1914 MOREAU, F. Sur le développement du périthèce chez une Hypocreale le *Peckiella laterita*, (Fries) Maire, R. Bull. Soc. Bot. de France, lxi, p. 160.

### *Hypocreaceae*

*Polystigma* is a small genus, the best-known member of which, *P. rubrum*, develops on the leaves of *Prunus spinosa*, of *P. insititia* and of the cultivated plum, where it produces conspicuous orange, yellow or scarlet stromata. Each of these is the result of a separate infection, and spreads over only a small part of the leaf, so that in autumn, when the leaves are shed, the host is freed from the disease. The fungus, however, hibernates in the fallen leaves, and next spring the ascospores mature reach the young leaves and there germinate.

Its development was first studied by Fisch in 1882, and by Frank in 1883, and these authors described trichogynes and the union of the latter with spermatia. More recent investigations, however, have shown that these organs, though present, are now no longer functional.

The germinating ascospore gives rise to a mycelium which ramifies among the cells of the host and forces them apart; the hyphae become massed especially in the intercellular spaces below the stomata, and often push their way to the exterior between the guard cells. Finally the stroma may extend from the upper to the lower epidermis, and only a few isolated cells of the mesophyll remain in the infected region. The hyphae are multinucleate, they contain orange pigment and their originally thin walls are modified to form thick gelatinous membranes perforated by fine pits. The gelatinous walls are probably utilized as reserve material, for they are partly absorbed during the later stages of development after the fall of the leaf.

During the summer, large flask-shaped spermogonia appear and open on the underside of the leaf. usually in the position of a stoma. The wall of the spermogonium consists of densely interwoven filaments and it is lined by thin, uninucleate spermatial hyphae (fig. 106). The mature spermatium is a filiform curved structure, narrowed at its free end; it contains a single, much elongated nucleus, staining homogeneously, and occupying the lower half or two-thirds of the cell. All attempts to bring about the germination of these spermatia have failed, and no relation of any kind has been demonstrated between them and the female organ, consequently they must be regarded as no longer functional, and their original use can be inferred only from their structure. Their small size, scanty contents, and large nucleus suggest that they are more appropriately constituted to act as fertilizing agents than as a means of vegetative propagation.

The archicarp first appears as a multinucleate hypha, which becomes septate and somewhat elaborately coiled. The base can usually be traced to a vegetative filament; the apex ends freely in the mass of uninucleate mycelial cells (fig. 102); most of the cells of the archicarp contain several nuclei, but a few are uninucleate. The archicarps usually develop singly, generally below or near a stoma, through which vegetative filaments project (fig. 107). These projecting hyphae were regarded by Fisch and Frank as

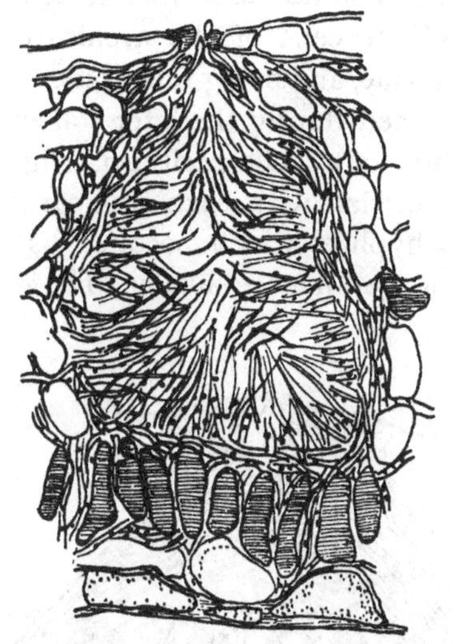

Fig. 106. *Polystigma rubrum* DC.; spermogonium, × 250; after Blackman and Welsford.

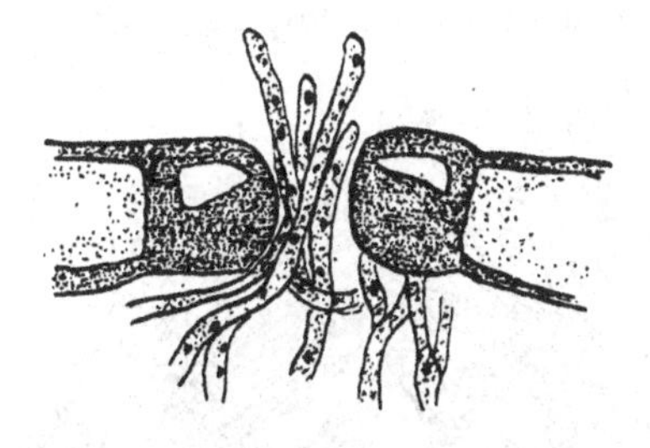

Fig. 107. *Polystigma rubrum* DC.; vegetative hyphae projecting through stoma above archicarp, × 900; after Blackman and Welsford.

trichogynes, but Blackman and Welsford, and later Nienburg, failed to trace any connection between them and the coiled archicarps. On the contrary, the latter end blindly within the stroma with or without branching, and it is only quite occasionally that they can even be traced upwards towards the stomata.

Nienburg observed the formation of a pore between a multinucleate cell at the base of the archicarp and the large uninucleate cell next in order to it. At a later stage he found that the uninucleate cell had become binucleate, the nuclei being at first somewhat different in structure, and that certain large cells, which apparently developed from it, also contained two nuclei each. In his opinion, the second nucleus in the originally uninucleate cell, is derived from its multinucleate neighbour, which he terms the antheridium; the other binucleate cells receive their nuclei from it by conjugate division, and are the beginnings of ascogenous hyphae. Though he was unable to see either the entrance of the second nucleus, or the process of conjugate division, his facts are decidedly suggestive, but they point less to normal fertilization than to the pseudapogamous association of a vegetative and a female nucleus.

The binucleate character of the later formed large cells may, as he suggests, be due to conjugate division, but, since he finds that the numerous binucleate cells in the sheath[1] are the result of rapid growth, this character in the large cells is evidently susceptible of the same explanation. In any case the rest of the archicarp degenerates and owing to the refractory character of the material the ascogenous hyphae could not be further traced.

According to Blackman and Welsford, all the cells of the archicarp degenerate without giving rise to ascogenous hyphae, and being functionless, retain their contents so that they can be recognized during the later stages of development as densely staining masses (fig. 108). The perithecia (fig. 109) arise in their neighbourhood, one in association with each archicarp, and the vegetative cells produce ascogenous hyphae, which become distinguished by their large size, dense contents and well-marked nuclei. These

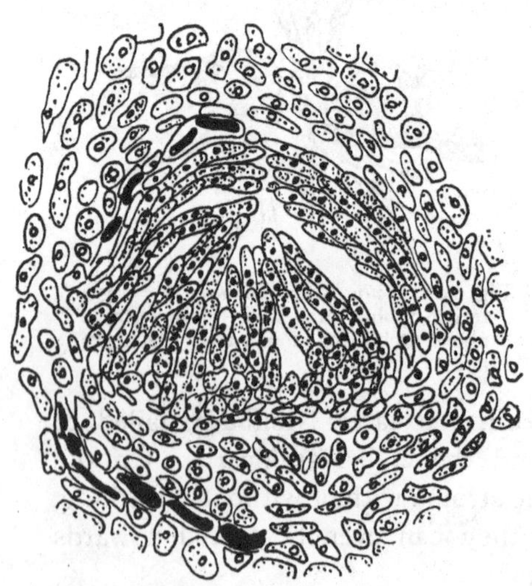

Fig. 108. *Polystigma rubrum* DC.; young perithecium; the ascogenous hyphae are not yet clearly distinguished, many of the nuclei are in pairs, the darkly stained remains of the archicarp are visible near the periphery; × 680; after Blackman and Welsford.

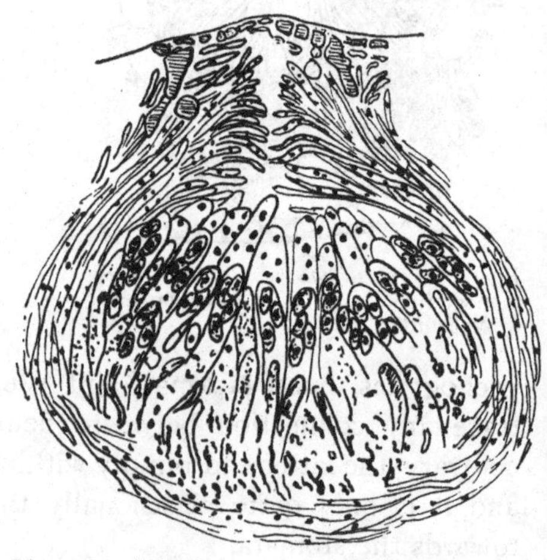

Fig. 109. *Polystigma rubrum* DC.; mature perithecium, × 270; after Blackman and Welsford.

[1] Nienburg, p. 390, end of first paragraph.

authors found some evidence that a first nuclear fusion takes place in the ascogenous hypha before the differentiation of the asci.

The ascus is formed in the usual way from the penultimate cell of the hypha; the usual nuclear fusion and successive nuclear divisions take place during its development.

In the genus *Podocrea*, the stroma is erect, and sometimes branched; in *Hypocrea* it is usually hemispherical or bolster-shaped and is colourless, or yellow or brown in colour. The majority of species occur saprophytically on wood, or as parasites on the larger fungi. In both genera and in their immediate allies, the spores are two or more celled. The systematic position of *Podocrea alutacea*[1] has undergone curious vicissitudes; in consideration of its form it was at first placed among the Basidiomycetes as *Clavaria simplex*, later it was regarded as a compound structure, the pyrenomycetous fungus being held to be parasitic, according to different authors, on *Clavaria ligula* and on species of *Spathularia*. The stalk being thus attributed to another fungus, the ovoid perithecial portion was referred to the genus *Hypocrea*. The question was set at rest by Atkinson, who succeeded in growing the normal upright stromata in pure culture from ascospores alone, and thus demonstrated that only one fungus was concerned.

The species of *Epichloë* occur parasitically on grasses the stems of which become coated by the stromata. The stroma is at first white, then yellow; in the early stages of its development oval conidia are produced, later the perithecia, which are completely embedded in the stroma, reach maturity; the ascospores, like those of the remaining genera of the Hypocreaceae, are filiform; Dangeard has shown that they are at first elliptical and uninucleate; later they elongate, the nucleus divides and the spore undergoes septation.

The genus *Cordyceps* (fig. 110) includes about sixty species; these are mainly tropical forms parasitic on insects, the bodies of which they transform into sclerotia from which the stromata grow out. The peculiar appearance of these structures has given rise to curious views as to their significance and medicinal value; thus Berkeley reports that *Cordyceps sinensis* is a "celebrated drug in the Chinese pharmacopoeia, but from its rarity only used by the Emperor's physician." The striking belief that it is "a herb in summer and a worm in winter," may perhaps sufficiently account for the esteem in which it was held.

The ascospores are multicellular and filiform and when shed break up into their separate cells. Germ-tubes from these, or from the conidia, infect the insect either as a caterpillar or chrysalis, and penetrating into its interior give rise to cylindrical conidia which enter the blood-stream and increase by yeast-like budding till the insect dies. A mycelium then appears and

[1] *Podocrea alutacea* Lindau = *Podostroma alutaceum* (Pers.) Atkinson.

the formation of the sclerotium begins; chains of subaerial conidia may be produced on conidiophores arranged in a coremium or fascicle of parallel hyphae. This is the *Isaria* condition, and though there is little doubt that it is a stage in the development of the *Cordyceps*, the ultimate proof by culture has yet to be given.

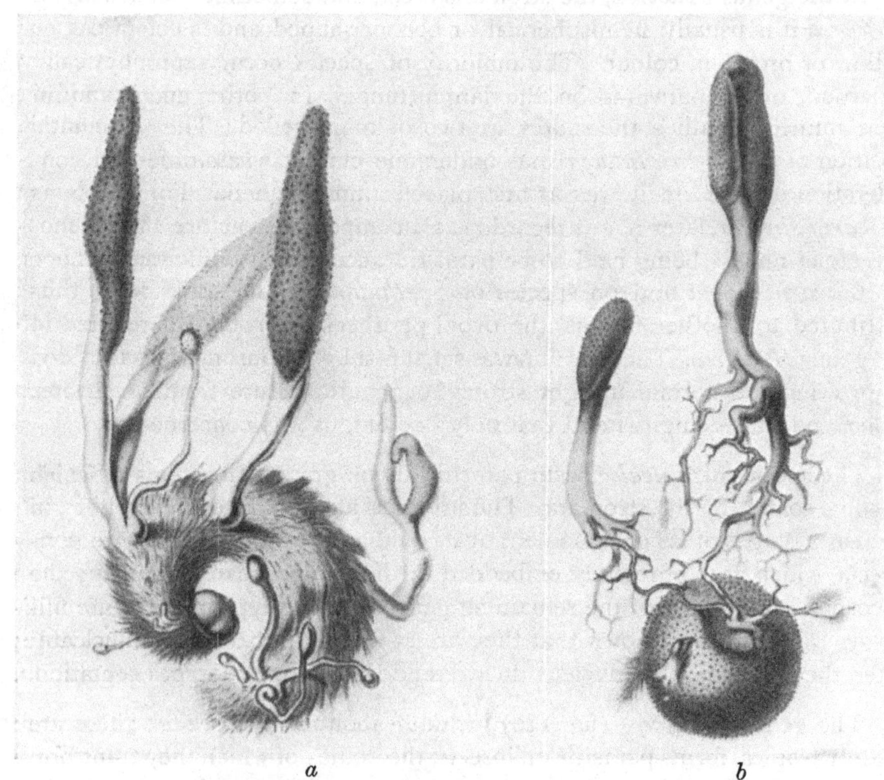


Fig. 110. *a. Cordyceps militaris* (L.) Link; *b. Cordyceps ophioglossoides* (Ehrh.) Link; after Tulasne.

The mature sclerotium is a compact mass of interwoven hyphae whose cells are rich in glycogen and oily matter. During its development the internal organs of the host are completely destroyed and replaced by the mycelium, the skin alone remaining intact. From this mummified structure one or more stromata arise, emerging between two segments of the skin, usually near the head. The stroma is differentiated into an erect, sterile stem, which may be simple or branched, and a globose or elongated, fleshy, fertile portion, usually terminal on the stem and bearing the perithecia (fig. 111). It is pale or bright coloured; red in the best known British species, *C. militaris*; and in other forms, purple, flesh-coloured, lemon-yellow or of various shades of brown.

As development proceeds the ovate or flask-shaped perithecia are differentiated; they always arise deep in the stroma and may remain completely or partially immersed or may become superficial as they approach maturity. Where they are more or less free the surface of the head is usually rough, whereas when they are entirely immersed it is smooth, but in some cases the free perithecia stand so close together as to produce a smooth appearance. The cytological details of development have not been studied; the perithecia arise from the vegetative cells of the stroma and in no case have any signs of sexual organs been seen; it would thus appear that *Cordyceps* is completely apogamous. The first sign of the perithecium is the differentiation of a knot of deeply staining vegetative hyphae.

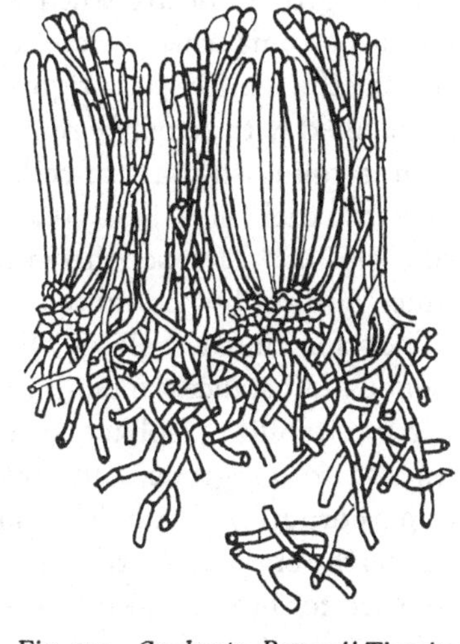
Fig. 111. *Cordyceps Barnesii* Thwaites; perithecia, × 170; after Massee.

The asci are long and slender with slightly swollen apices into which the spores do not penetrate; at maturity the contents of the apex swell and the wall is ruptured. The spores are arranged in a parallel manner, in a fascicle slightly twisted on its axis, and are nearly as long as the ascus; they are hyaline, very slender and almost always multicellular; they break up readily into their constituent cells which, as already stated, germinate separately to infect a new host. According to the investigations of Lewton-Brain several nuclear divisions take place in the ascus before spore-formation and the spores are multinucleate from their first inception.

Two species, *C. ophioglossoides* (fig. 110 *b*) and *C. capitata*, are parasitic on underground fungi of the genus *Elaphomyces* and do not produce true sclerotia; for these reasons they are sometimes separated as another genus *Cordylia*.

The species of *Claviceps*, like those of *Cordyceps*, possess filiform ascospores, and form sclerotia from which the stromata arise. The genus is, however, much smaller, including only six species parasitic on various Gramineae. Of these the best known is the almost cosmopolitan species *C. purpurea*, the ergot, on rye and other cultivated grasses.

The ascospores germinate on the flowers of the host, and give rise to a mycelium which ramifies at first in the outer coats of the ovary and ultimately fills its whole cavity, forming a sclerotium. Outside the ovary, conidia are budded off, and at the same time a sweet fluid, the so-called honey-dew, is excreted; it attracts insects which carry the conidia with them to other flowers, where they at once germinate, and further infections are produced. On the completion of the conidial stage the sclerotia assume a firmer texture, and become dark purple or bluish black in colour. If they fall to the ground or are sown with the seed they give rise next spring to numerous stromata with violet stalks and reddish yellow heads. According to Fisch, the perithecia originate from two or three hyphal cells, which become filled with strongly refractive protoplasm and divide in all directions to form a roundish mass of cells distinguished from those of the rest of the stroma by their size and contents. As in *Cordyceps*, there is no trace of sexual organs. The perithecia are immersed in the stroma, and the asci produce filiform but continuous spores.

The sclerotium is well supplied with reserve materials and contains certain poisonous substances including ergotic acid, a narcotic which diminishes reflex excitability; sphacelic acid, the main cause of ergot poisoning, it gives rise to gangrene, and large doses produce tetanus of the uterus and cramp; cornutin, an alkaloid causing contraction of the uterus.

Thus the ergot sclerotia, if eaten with the grass by cattle, or included in the grain used for bread-making, are responsible for serious disease. When grain was less carefully purified than at present the inhabitants of whole districts sometimes became afflicted with gangrene, and the occurrence of the sclerotia in pastures is liable, owing to the presence of cornutin, to cause abortion in sheep or cows, so that many local traditions as to the prevalence of abortion in certain farms, or in certain byres, are probably traceable to this cause.

Cornutin is of medicinal value, and the sclerotia are collected for this purpose.

## HYPOCREACEAE: BIBLIOGRAPHY

1843 BERKELEY, M. J. On some Entomogenous Sphaeriae. London Journal of Botany, ii, p. 205.

1882 FISCH, C. Beiträge zur Entwickelungsgeschichte einiger Ascomyceten. Bot. Zeit. xl, p. 850.

1883 FRANK, E. Ueber einige neue oder wenige bekannte Pflanzen Krankheiten. Ber. der deutsch. bot. Ges. i, p. 761.

1895 MASSEE, G. A Revision of the Genus Cordyceps. Ann. Bot. ii, p. 207.

1901 LEWTON-BRAIN, L. *Cordyceps ophioglossoides*. Ann. Bot. xv, p. 522.

1905 ATKINSON, G. F. Life-history of *Hypocrea alutacea*. Bot. Gazette, xl, p. 401.

1907 DANGEARD, P. A. Recherches sur le développement du périthèce chez les Ascomycètes. Le Botaniste, x, p. 352.

1912 BLACKMAN, V. H. and WELSFORD, E. J. The Development of the Perithecium of *Polystigma rubrum*. Ann. Bot. xxvi, p. 761.

1914 NIENBERG, W. Zur Entwickelungsgeschichte von *Polystigma rubra*. Zeitschr. für Bot. vi, p. 369.

## DOTHIDEALES

The Dothideales constitute a small group of some four hundred species, included in twenty-four genera, forming a single family, the Dothideaceae. They are parasites or saprophytes on the leaves and stems of higher plants, on which they produce stromata usually below the epidermis and finally exposed by its rupture.

The stroma is externally black and hard, built up of hyphae closely interwoven to form a pseudoparenchyma; internally it is of much looser consistency, and is often white or brownish in colour. The perithecia are without definite walls, so that the asci develop in mere cavities in the stroma, which however have the globose form of ordinary perithecia, and are bordered by cells rather smaller and narrower than those of the surrounding mycelium. In some cases, where the inner tissue of the stroma is very loosely interwoven, the perithecium is, however, definitely delimited.

In *Dothidea* the stromata form black projecting cushions, which in *D. virgultorum* occur on the living, as well as the dead stems and branches of the birch.

In *Plowrightia* the very similar stromata run together in masses. *P. morbosa* is a serious disease attacking species of *Prunus*, especially the cherry and plum. The mycelium penetrates the living branches which become swollen and deformed and on which stromata and finally perithecia are produced.

## Sphaeriales

The Sphaeriales are distinguished by the dark colour and membranous, corky or carbonaceous texture of their perithecia, and of their stromata when present. They number already considerably over six thousand species, and new species are constantly being brought to light, so that there is no doubt that a study of the tropical forms, at present very incompletely known, will greatly increase their number. Not only the number of species, but the number also of individuals is very considerable; the majority are saprophytes, and serve a useful purpose in bringing about the first stages of decay in such resistant materials as wood and straw. They greatly outnumber the Hypocreales and Dothideales, and it is from their black or brown colour and often charred appearance that the name Pyrenomycetes is derived.

The origin of the group has been proposed through *Chaetomium*, which is sometimes without an ostiole, from the Erysiphales, or, in view of the structure of the sexual organs, from an *Eurotium*-like form among the Plectascales.

Unfortunately their small size and resistant texture as well as the nature of their habitat make many of the simpler species unfavourable subjects of study, and our knowledge of their development is at present very fragmentary. *Sordaria* and some others can be grown on artificial media and satisfactory results may be anticipated from a further application of this method. Some of the larger forms with a well-developed stroma can readily be handled but in none of these has normal sexuality yet been observed.

The perithecia in the simplest forms are borne singly, free or partially embedded in the substratum; from these may be traced a series of intermediate forms culminating in the elaborate stromata and sunken perithecia of the highest species. There is, in fact, a marked parallelism between the Sphaeriales and Hypocreales, and it is by no means clear that the colour and texture of the stroma and perithecium walls are of sufficient importance as criteria of relationship to justify their separation, nor is it indicated that the members of the families Nectriaceae or Hypocreaceae resemble one another more closely than the numerous Sphaeriales, though these are dispersed among a series of eighteen or nineteen families. The method of classification is however convenient, and considerably more knowledge will be required before a natural system of classification can be elaborated. In the meantime, the subdivisions of the Sphaeriales rest on the structure and development of the stroma, the form of the ostiole, and the colour and septation of the spores. As in the Hypocreales, various sorts of accessory fructifications are present.

In the first eight families of the Sphaeriales the perithecia are more or less free, though they may be partly sunk in the substratum, or in a weft of hyphae, or may be seated on a definite stroma.

In the remaining ten families the perithecia are immersed either in the substratum, or in a stroma which may reach considerable elaboration.

The most important of the eighteen families of the Sphaeriales are:

| | |
|---|---|
| Perithecia free | |
| Peridium membranous | |
| ostiole beset with long hairs often elaborately coiled or branched | CHAETOMIACEAE. |
| ostiole without long hairs; mainly coprophilous | SORDARIACEAE. |
| Peridium leathery or carbonaceous | |
| short neck | SPHAERIACEAE. |
| long, sometimes filiform neck | CERATOSTOMATACEAE. |
| Perithecia embedded in substratum | |
| Perithecia immersed, upper part free | |
| ostiole round | AMPHISPHAERIACEAE. |
| ostiole elliptical | LOPHIOSTOMATACEAE. |
| Perithecia completely immersed, ostiole only projecting | |
| peridium membranous or leathery, neck short | |
| paraphyses absent | MYCOSPHAERELLACEAE. |
| paraphyses present | PLEOSPORACEAE. |
| peridium leathery or carbonaceous, neck long | GNOMONIACEAE. |
| Perithecia embedded in stroma | |
| Stroma developed within substratum, differentiated from it | VALSACEAE. |
| Stroma free | |
| ascospores very small, sausage-shaped and hyaline or light brown, unicellular | DIATRYPACEAE. |
| ascospores unicellular, rarely bicellular, dark brown | XYLARIACEAE. |

*Chaetomiaceae*

The Chaetomiaceae occur on straw, paper, dung and other waste materials; they possess free, thin-walled perithecia beset with numerous characteristic, long hairs (fig. 112), which are often elaborately branched or coiled. On these, or on the ordinary vegetative mycelium, conidia are produced. An ostiole is lacking in *Ch. fimete*, presumably the most primitive member of the genus; in the remaining species it is present and the perithecium is of the typical sphaeriaceous form.

In *Chaetomium spirale* the cells of the mycelium contain each a single nucleus, the archicarp arises as a coiled branch and divides into four or more uninucleate cells. There is no sign of an antheridium. Vegetative hyphae grow up from the stalk of the archicarp, and from the filament on which it is borne, and form a sheath, the outer cells of which are prolonged as hairs. Small pyriform conidia are abundant.

*Ch. Kunzeanum* shows a very similar archicarp (fig. 113), but here the cells, as described by Vallory for the variety *chlorinum*, each contain several nuclei which are often found approximated in pairs. This arrangement is reported to be as common in the vegetative mycelium as in the cells of the archicarp, and is doubtless a result of rapid division.

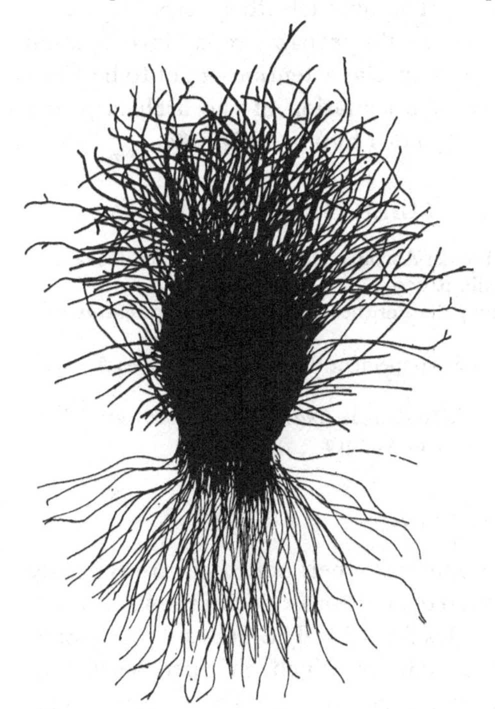

Fig. 112. *Chaetomium pannosum* Wallr.; × 50; W. Page del.

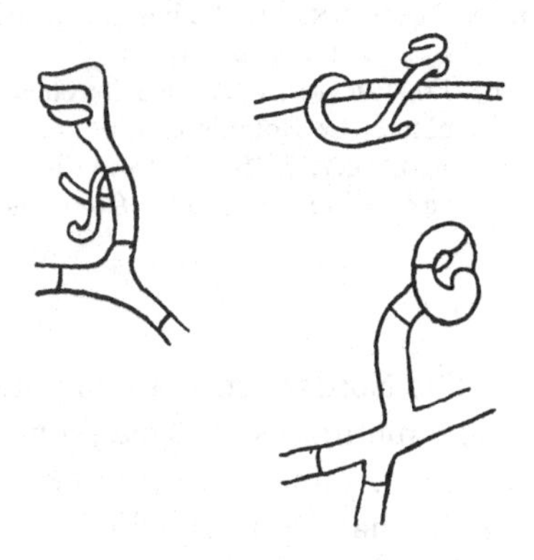

Fig. 113. *Chaetomium Kunzeanum* Zopf; archicarps; after Oltmanns.

In due course the archicarp becomes surrounded by a sheath of vegetative hyphae within which its growth is continued so that a mass of cells is produced from which asci at last arise. In the meantime the sheath becomes differentiated into an outer coat of relatively large, brown-walled hyphae, and an inner layer of smaller cells which become narrow and elongated. As development proceeds a cavity appears within the perithecium, usually just above the ascogenous cells, and branches from the lining mycelium grow out to form the periphyses; paraphyses are not produced (fig. 114).

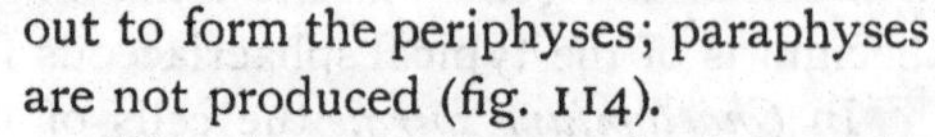

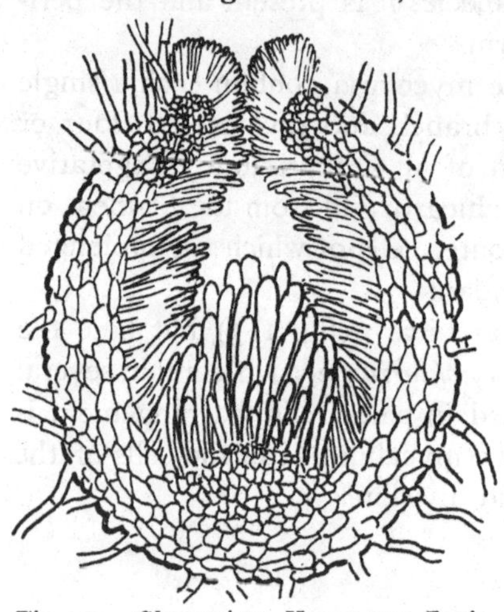

Fig. 114. *Chaetomium Kunzeanum* Zopf; perithecium, × 200; after Zopf.

The ripe spores are shed into the cavity of the perithecium, and do not reach the exterior immediately on leaving the ascus.

In addition to the above, two or three other species have been examined, and show the same type of archicarp and of perithecium, but in no case has any further cytological detail been worked out.

The uninucleate species in particular would probably repay investigation and special attention ought to be given to the septation of the archicarp and to the number of cells from which ascogenous hyphae originate.

## CHAETOMIACEAE: BIBLIOGRAPHY

1881 ZOPF, W. Zur Entwickelungsgeschichte der Ascomyceten. Chaetomium. Nova Acta Acad. C. Leop.-Carol. G. Nat. Cur. xlii, p. 199.

1887 OLTMANNS, F. Ueber die Entwickelung der Perithecien in der Gattung *Chaetomium*. Bot. Zeit. xlv, p. 193.

1907 DANGEARD, P. A. Recherches sur le développement du périthèce chez les Ascomycètes. Le Botaniste, x, p. 329.

1911 VALLORY, J. Sur la formation du périthèce dans le *Chaetomium Kunzeanum* Zopf. var. *chlorinum* Mich. Comptes Rendus, cliii, p. 1012.

### *Sordariaceae*

The Sordariaceae are mainly coprophilous; their perithecia are typically free, sometimes superficial, sometimes so deeply embedded in the substratum that little more than the neck protrudes from it. The genus *Hypocopra* is exceptional in possessing a small stroma in which the perithecium is immersed, but it resembles *Sordaria* in all other points. The present family differs from the Chaetomiaceae in bearing only short filaments instead of long hairs around the ostiole, and from the Sphaeriaceae in the habitat and type of spore. The mycelium is in most cases composed of multinucleate cells, but in *Podospora hirsuta* the cells are uninucleate (fig. 115), recalling the condition in several species of *Chaetomium*.

The commonest type of archicarp is a stout, coiled, septate hypha which soon becomes surrounded by vegetative filaments; it is usually terminal, but is occasionally intercalary, for instance in *Sordaria fimicola*. Dangeard has found a straight archicarp (fig. 116) in *Sordaria macrospora*, and in 1868, for *S. fimiseda*, Woronin described an archicarp with a swollen terminal cell recalling the oogonium of *Humaria granulata*.

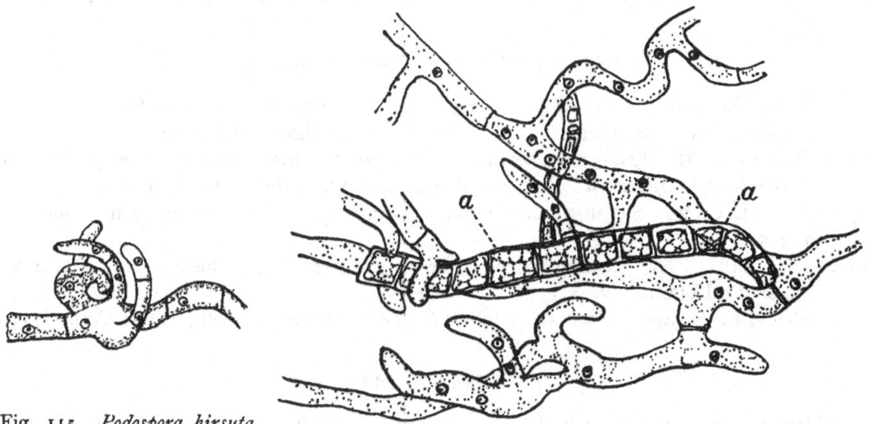


Fig. 115. *Podospora hirsuta* Dang., archicarp; after Dangeard.

Fig. 116. *Sordaria macrospora* Auersw.; *a*. straight archicarp; after Dangeard.

In *Sporormia intermedia* the perithecium is initiated by the enlargement of a multinucleate mycelial cell which is often intercalary. It undergoes not only transverse but also longitudinal divisions, forming a pseudoparenchymatous mass of uninucleate cells (fig. 117), with which various neighbouring cells anastomose. The mass thus formed is responsible for the whole contents of the perithecium, though the outer walls may be formed by ordinary vegetative hyphae. In view of this fact it seems doubtful whether the initial cell should here be regarded as an oogonium, that is to say as having at one time had a sexual significance, and not rather as a preliminary stage in the development of such a mass of hyphae as initiates the apogamous perithecium of *Claviceps* and its allies.

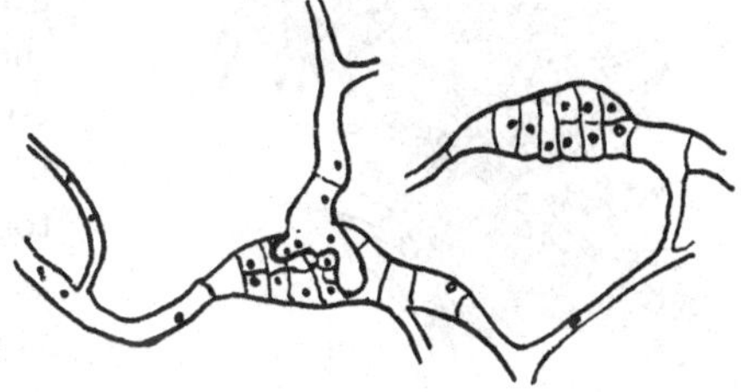

Fig. 117. *Sporormia intermedia* Auersw.; initial cells of perithecium; after Dangeard.

In some of the Sordariaceae each spore is surrounded by a layer of mucilage (*Sordaria macrospora*, *S. fimicola*, etc.), in others (fig. 2 *e*) one or two appendages are produced (*S. fimiseda*, *S. coprophila*, *Podospora anserina*, etc.). These may be gelatinous and derived wholly or partly from the epiplasm apparently much as the ordinary thickening of the spore wall is derived, or they may form part of the young spore. In the latter case they are at first rich in protoplasm, but later most of their contents pass into the middle portion of the spore, which becomes ovoid, and the appendage is cut off by a wall (*S. globosa*). Both types of appendage may occur on the same spore. They are sometimes hooked and they become twisted together and serve to attach the spores one to another. The uppermost appendage appears, in some cases at any rate, to become fastened to the tip of the ascus (*S. Brefeldii*.

## SORDARIACEAE: BIBLIOGRAPHY

1883 ZOPF, W. Zur Kenntniss der anatomischen Anpassung der Pilzefruchte an die Funktion der Sporentleerung. Zeitschr. für Naturwiss. iv; ii, p. 540.

1886 WORONIN, M. *Sphaeria Lemaneae*, *Sordaria fimiseda*, *Sordaria coprophila* und *Arthrobotrys oligospora*. Beit. zur Morph. und Phys. der Pilze, iii, p. 325.

1901 MASSEE, G., and SALMON, E. Researches on Coprophilous Fungi. Ann. Bot. xv, p. 315.

1907 DANGEARD, P. A. Recherches sur le développement du périthèce chez les Ascomycètes. Le Botaniste, x, p. 333.

1912 WOLF, F. A. Spore Formation in *Podospora anserina*, (Rabh.) Wint. Ann. Myc. x, p. 60.

### *Sphaeriaceae*

The perithecia of the Sphaeriaceae are superficial, and are borne singly or in groups; the peridium may be smooth or beset with hairs or spines. The papillate ostiole distinguishes this family from the succeeding forms with free perithecia.

The majority are saprophytic on plant remains, frequently on wood; some are parasites, such as the species of *Coleroa* (fig. 118), which occur on the leaves of *Potentilla*, *Rubus*, and other flowering plants.

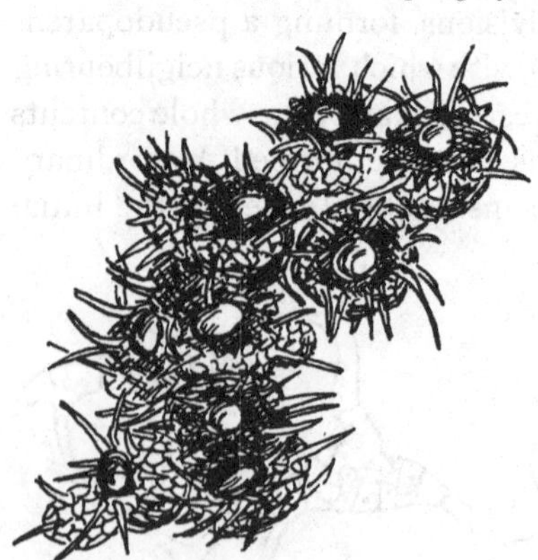

Fig. 118. *Coleroa Potentillae* (Fr.) Wint.; perithecia, × 192.

*Rosellina quercina*, the oak root fungus, attacks the roots of oak seedlings; the hyphae enter the living cells of the cortex and pith; they are at first hyaline, later dark in colour, and become twisted together into strands, the so-called rhizoctonia; these attack the roots of neighbouring oak plants, wrap a weft of hyphae about them and enter their cells. The fungus may form black, chambered sclerotia which originate in the cortex of the host root; reproduction is by means of conidia formed in summer on the surface of the soil, and further by ascospores produced in perithecia. Hartig has found that the perithecium is initiated by the development of a pair of thick hyphae rich in contents. These become enclosed within a mass of vegetative tissue, but their subsequent behaviour has not been determined, and no details of development are known either here or in other members of the family.

## SPHAERIACEAE: BIBLIOGRAPHY

1880 HARTIG, R. Der Eichenwurzeltodter *Rosellina quercina.* Untersuch. aus der forstbotanische Inst. zu München iii, p. 1.

### *Ceratostomataceae*

The Ceratostomataceae resemble the Sphaeriaceae in most of their characters; they are distinguished by the elongated neck of the perithecium, which is often drawn out to form a delicate hair-like process. The method of liberation of the spores in this family presents an interesting problem, but neither that question nor the development of the perithecium has yet been elucidated.

### *Amphisphaeriaceae*

In the Amphisphaeriaceae the young perithecium is sunk in the substratum; as it matures it becomes more or less free, though in contrast to the condition in the Sphaeriaceae and Ceratostomataceae, its base is always immersed.

Development has been studied in a species of *Teichospora* and a species of *Teichosporella*, now both included under the genus *Strickeria*, characterized by its muriform spores. The spore produces numerous germ-tubes which give rise to a mycelium of multinucleate cells; certain cells increase in size and become both transversely and longitudinally divided till a parenchymatous mass is produced (fig. 119). Other vegetative hyphae may form a scanty investment, but often the perithecium develops without this addition.

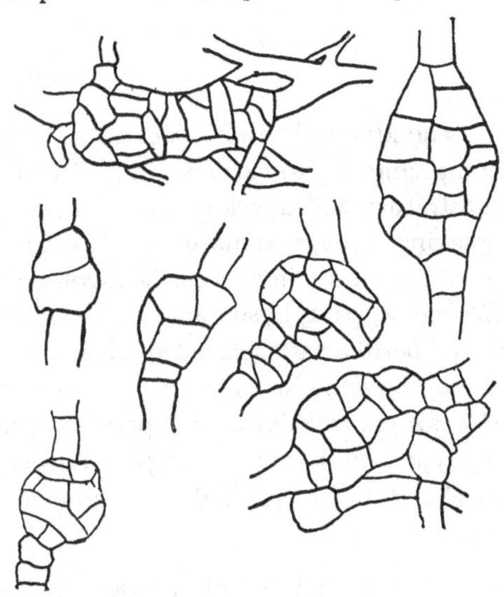

Fig. 119. *Strickeria sp.*; initial cells of ascocarps; after Nichols.

Asci appear as large uninucleate cells, and the tissue around them disorganizes. The outer hyphae become hard and dark only when the perithecium approaches maturity.

The type of development here is very similar to that already described for *Sporormia*; it seems very doubtful whether the initial cell of the perithecium should be regarded as an oogonium, or whether the development is purely vegetative. The peculiarity in either case is the formation of the bulk of the perithecium from a single cell instead of, as in the majority of forms, from a complex of interwoven hyphae differentiated into sexual and vegetative components.

## AMPHISPHAERIACEAE: BIBLIOGRAPHY

1896 NICHOLS, M. A. The Morphology and Development of certain Pyrenomycetous Fungi. Bot. Gaz. xxii, p. 301.

### *Lophiostomataceae*

The perithecia of the Lophiostomataceae are borne singly; during development they are embedded in the substratum, and they may so remain or may become partially free at maturity. There is no stroma, and the peridium is black and brittle. So far there is a close resemblance to the Amphisphaeriaceae, but the Lophiostomataceae are distinguished by the form of the ostiole, which is very large and laterally compressed, so that in external appearance they approach certain of the Hysteriales which in many cases they further resemble in their habitat on vegetable remains such as wood and bark. None of the species has been investigated in detail.

### *Mycosphaerellaceae*

The Mycosphaerellaceae are parasitic forms occurring usually on leaves and giving rise to various kinds of leaf-spot. The perithecia are sunk in the substratum and develop either under the cuticle or beneath the epidermis, breaking through at maturity. The ascospores are usually septate, frequently bicellular and sometimes dark-coloured; except in the transitional genus *Stigmatea*, paraphyses are not developed. In several cases the formation of the perithecia is preceded by a conidial stage.

*Mycosphaerella nigerristigma* forms pycnidia on the living leaves of *Prunus pennsylvanica* and perithecia after the leaves have fallen. A trichogyne like that of *Polystigma* has been recorded; it degenerates, leaving a basal cell, but whether this functions is not known.

## MYCOSPHAERELLACEAE: BIBLIOGRAPHY

1914 HIGGINS, B. B. Life-History of a new species of *Sphaerella*. Myc. Centralbl. iv, p. 187.

### *Pleosporaceae*

The Pleosporaceae are saprophytes or in a few cases parasites, for the most part on seed plants but in some cases on Pteridophyta, Bryophyta or Lichens. The perithecia are immersed in the substratum, the ostiole only projecting, but they may become more or less exposed by the rupture of the covering tissues. The peridium is leathery or membranous.

The genus *Pleospora* includes some 225 species, several of which occur on grains and other grasses where they show biological specialization. *Pleospora herbarum* is a facultative parasite on the leaves of angiosperms; the perithecium is initiated by the division of a hypha into numerous short cells from which branches grow out. The central cells, and later the basal parts of the branches, divide in various directions till an irregular parenchymatous mass is formed. By further growth and division the mass assumes a globular shape and the central cells become elongated and differentiated as paraphyses. Later, asci appear, developing from the same cells as the paraphyses and each produces eight muriform spores (fig. 120).

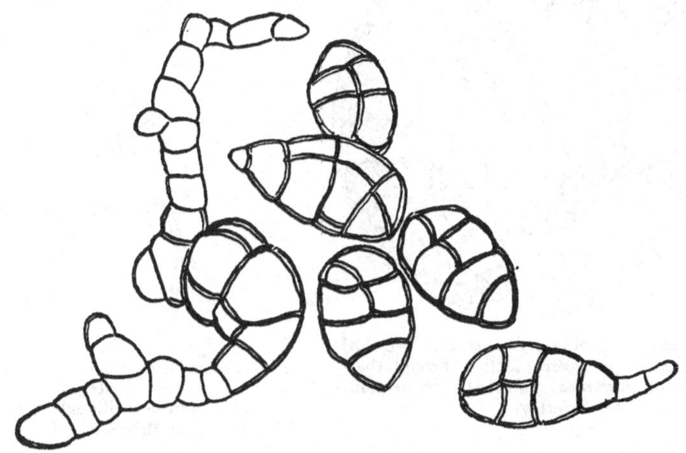

Fig. 120. *Pleospora* sp.; germinating spores, × 1000.

Multicellular conidia also develop on branched hyphae, the terminal cells of which form the sterigmata. After the spore is shed the hypha may continue to grow, a new sterigma being formed above the old one. The name *Macrosporium parasiticum* was formerly applied to the conidial stage of this species.

The genus *Venturia* includes over fifty species, several of which are parasitic on living leaves; the perithecium is immersed and the large ostiole beset with stiff hairs or bristles. The species grouped under *Fusicladium* among the Hyphomycetes are in some cases conidial forms of this genus. The conidia are two-celled, borne on short conidiophores arranged in groups; *F. dendriticum* is the cause of scab or black-spot on apples, and *F. Pyrinum* of a similar disease on pears.

*Leptosphaeria* includes some 500 species characterized by the papillate or conical ostiole, usually free from hairs. The majority are saprophytes on plant remains, some are parasites on land plants, and some on the Red Algae.

*L. Lemaneae* occurs on the thallus of various species of *Lemanea* (fig. 121). The mycelium consists of uninucleate cells and ramifies in the intercellular spaces of the host, sending branched haustoria into the cells. Here and there the hyphae are dilated (fig. 122 *a*, *b*), and in these regions show denser and more refractive contents than usual. Fusion takes place between the dilated portions (fig. 122 *c*, *d*) which may be terminal or intercalary, and there is

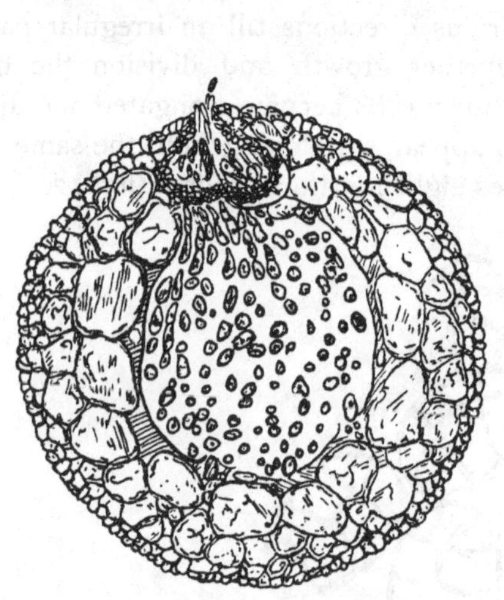

Fig. 121. *Leptosphaeria Lemaneae* (Cohn) Brierley; transverse section through thallus of *Lemanea*, showing perithecium, × 125; after Brierley.

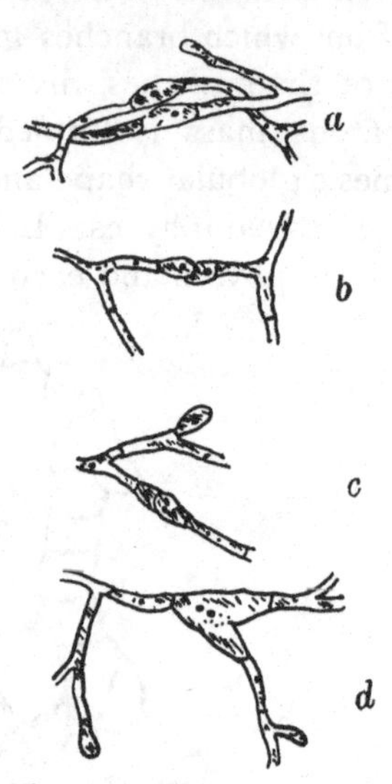


Fig. 122. *Leptosphaeria Lemaneae* (Cohn) Brierley; *a*. *b*. *c*. *d*. stages of fusion between dilated hyphae; after Brierley.

evidence that the nucleus of one of the swollen cells passes across into the other, which may therefore be termed the oogonium, and fuses with its nucleus. The oogonium then divides to form a number of multinucleate cells from which ascogenous hyphae arise. The nuclei in these hyphae are paired and the usual fusion takes place in the ascus. From the cells adjoining the oogonium the delicate hyphae of the sheath grow up. The morphology of the sexual organs in this genus is quite unusual, but they may perhaps best be compared with the dilated cell observed by Dangeard in the initiation of the perithecium in *Sporormia intermedia*; in that case, however, there does not appear to be a functional antheridium, and vegetative cells as well as ascogenous hyphae are stated to develop from the initial cell; the resemblance demands further investigation.

The family is rich in conidial forms, and it is probable that several species of Fungi Imperfecti, including the pycnidial genera *Phoma* and *Hendersonia* and also *Cercospora*, a form with long septate conidia on free conidiophores are stages in the development of members of the Pleosporaceae.

## PLEOSPORACEAE: BIBLIOGRAPHY

1886 WORONIN, M. *Sphaeria Lemaneae*, *Sordaria fimiseda*, *Sordaria coprophila*, und *Arthrobotrys oligospora*. Beit. zur Morph. und Phys. der Pilze, iii, p. 325.

1889 MIYABE KINGO. On the Life History of *Macrosporium parasiticum*, Thüm. Ann. Bot. iii, p. 1.

1913 BRIERLEY, W. B. The Structure and Life History of *Leptosphaeria Lemaneae* (Cohn). Mem. and Proc. Manchester Lit. and Phil. Soc. lvii, 2, p. 1.

### *Gnomoniaceae*

The Gnomoniaceae are for the most part saprophytic on the leaves or other parts of plants. The perithecia are embedded in the substratum from which their long necks project. The ascus is characterized by a thickened apex through which a canal allows the exit of the spores. The spores are hyaline and paraphyses are usually not developed. The family differs from the Pleosporaceae in the long neck of the perithecium and the thickened apex of the ascus. There is no stroma, and this fact, as well as the dark colour, distinguishes *Gnomonia* from the similar genus *Polystigma* among the Hypocreales.

*Gnomonia erythrostoma* is the cause of an epidemic disease known as cherry-leaf-scorch, which attacks the foliage of *Prunus avium* and of several varieties of the cultivated sweet cherry. The mycelium ramifies on the leaf and runs back to the base of the petiole, where it prevents the formation of the absciss layer. In consequence the infected leaves do not fall, but remain hanging on the branches; they are the only source of infection in the following summer, and their destruction is therefore a sure method of checking the disease.

Infection usually takes place in June; towards the end of August spermogonia appear; they are shallower than those of *Polystigma*, but otherwise very like them, with a wall of closely compacted hyphae and a small circular ostiole opening on the under surface of the leaf. The spermatial hyphae are narrow and tapering, and their extremities are abstricted to form the spermatia, each of which contains a long threadlike nucleus and a relatively small amount of cytoplasm.

Soon after the spermogonia have begun to develop certain hyphae near the lower epidermis of the leaf become entwined to form more or less spherical coils, the primordia of the ascocarps. Their apices project in groups of four or five through the stomata, and the terminal cells become swollen and apparently mucilaginous; these projecting filaments were regarded by Frank as trichogynes, but more recently Brooks has found evidence that they arise from the outer cells of the perithecium and that, whatever their origin, they now no longer function as receptive structures. Spermatia are often found attached to their terminal cells, but, in view of the enormous number of spermatia liberated on the under surface of the leaf, they could hardly fail to be found in relation to any projecting filament.

In the lower part of the coils certain cells become differentiated by their denser cytoplasm and larger nuclei, and no doubt represent the oogonial regions of the archicarps. No union of nuclei has however been observed in them and it is at least doubtful whether they give rise to the ascogenous hyphae. The latter do not become clearly differentiated till the oogonial cells have disappeared; asci are formed either from the terminal or subterminal cells; in the young ascus two nuclei fuse.

Throughout the divisions in the ascus and in the division of the spore nucleus Brooks has reported four chromosomes. Those in the first division in the ascus are short and thick, resembling heterotype chromosomes in appearance, and there seems reason to believe that reduction occurs at this stage.

The life-history of *Gnomonia* shows many points in common with that of *Polystigma*; both are at first leaf parasites, and complete their development saprophytically on the dead leaf. Both produce spermogonia with filiform spermatia and perithecia developed in relation to coiled archicarps.

An important point of difference is that in *Polystigma* a stroma is formed and the fungus hibernates on the fallen leaves below the tree without being injured by their decay; in *Gnomonia* no stroma is present and the fungus inhibits the formation of the absciss layer so that the withered leaves remain on the branches and provide a matrix in which the perithecia can be formed.

## GNOMONIACEAE: BIBLIOGRAPHY

1886 FRANK, B. Ueber *Gnomonia erythrostoma*, die Ursache einer jetzt herrschenden Blattkrankheit der Süsskirschen im Altenlande, nebst Bemerkungen über Infection bei blattbewohnenden Ascomyceten der Bäume überhaupt, etc. Ber. der deutsch. Bot. Gesell. iv, p. 200

1910 BROOKS, F. T. The Development of *Gnomonia erythrostoma*, the Cherry-Leaf-Scorch Disease. Ann. Bot. xxiv, p. 585.

### *Valsaceae*

The perithecia of the Valsaceae are produced frequently in compact groups on a black stroma from which their long necks alone project. The stroma is very variable in form; it is developed within the substratum and more or less differentiated from it, sometimes indicated only by a black stain on the wood or bark of the host and by a black margin, sometimes extended as a thin black layer over a considerable area and ending irregularly; sometimes, as in species of *Valsa*, forming black cushions erumpent through the bark of the host. In a few cases the stroma surrounds only the upper part and not the base of the perithecium, and we have thus a transition from the rudimentary stromata of some of the earlier families.

The peridium is black and leathery, the asci usually long stalked, the spores uni- or multicellular, and hyaline or dark-coloured.

Conidia are frequently present, borne on free conidiophores or produced within pycnidia.

The genus *Valsa* includes some four hundred species and *Diaporthe* a rather larger number. The majority are saprophytic on wood and other resistant parts of plants.

### *Diatrypaceae*

In the Diatrypaceae the stroma is developed under the bark of the host, and forms either a cushion or a thin, flat layer which later becomes exposed. Conidia of various kinds are produced, but the conidial and perithecial stromata are often distinct and whereas the latter are of the usual dark colour and carbonaceous consistency the former are frequently light-coloured and fleshy. This separation and the usually unicellular, small, hyaline, curved ascospores are the principal characters of the family.

The genus *Calosphaeria* is exceptional in lacking a perithecial stroma; its perithecia are free and it could appropriately be placed in one of the groups near the Pleosporaceae but that a conidial stroma is present and closely resembles that of the Diatrypaceae; the ascospores, moreover, are of the characteristic curved form, so that *Calosphaeria* may, it appears, more fitly be regarded as a reduced member of the group. The species of *Calosphaeria*, like the other Diatrypaceae, occur chiefly on dead wood but *C. princeps* infects the living branches of cherry, plum and peach.

In *Diatrype* the most characteristic stroma is a black corky tissue of indefinite extent in which the perithecia are completely immersed; the ascus contains eight spores in contrast to the numerous spores of certain species of *Calosphaeria* and of *Diatrypella*, a genus further distinguished by the cushion-shaped stroma.

### *Xylariaceae*

The Xylariaceae occur chiefly on wood; they represent the highest development of the Sphaeriales and are characterized by the free superficial stroma which is only very rarely, as in *Hypoxylon*, partly sunk in the substratum, and shows every variety of form from a spreading crust on the surface of the host, as in the genus *Nummularia*, some species of which approximate *Diatrype*, to the almost spherical cushions of *Hypoxylon* (fig. 123) and the erect, simple, or branched stromata of *Xylaria* (fig. 124) and its allies. The perithecia are arranged just below and at right angles to the surface of the stroma; their development may be preceded by the formation of conidia which often cover the young stroma with a whitish powder.

Fig. 123. *Hypoxylon coccineum* Bull.; the smallest stroma bears conidia, the others perithecia; after Tulasne.

*Poronia punctata* occurs on old horse dung; the stromata are about 1 cm. in height, stalked and expanded above into a cup or disc (fig. 125), which, in the earlier stages of development, is covered by a greyish-white film of conidia; later the ostioles of the numerous perithecia appear as black dots scattered over the surface of the disc (fig. 126). The asci, when ripe, protrude through the ostiole so that the dark brown spores are shed outside the perithecium.

Dawson was able readily to obtain pure cultures, both from the ascospores and from the conidia, on 10 per cent. gelatine made up with decoction of horse dung.

The ascospore forms a single lateral germ-tube, which develops septa and branches freely. The conidia produce germ-tubes from either end or from both and sometimes also laterally; the mycelium is at first much more delicate than that derived from the ascospores but soon becomes indistinguishable from it. Branches arise from points just below the cross walls;

frequent lateral anastomoses occur and crystals of calcium oxalate, which have become separated from the substratum, are found among the filaments.

Hyphae become massed together to form the stroma which in the very young stages consists entirely of vegetative filaments densely inter-

Fig. 124. *Xylaria Hypoxylon* Grev., after Tulasne.

woven and rising vertically from the surface of the substratum. As they grow the stromata assume their characteristic shape, conidia appear and drops of pinkish or yellowish fluid are exuded. When these dry up, black dots indicating the position of the ripening perithecia are seen.

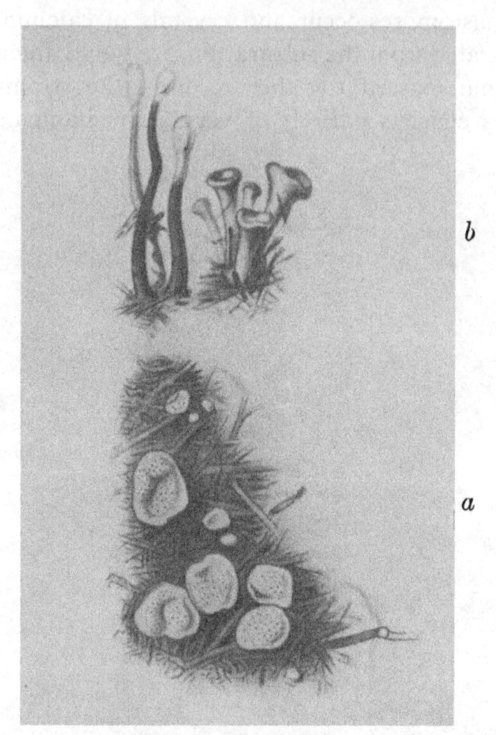


Fig. 125. *Poronia punctata* (L.) Fr.; *a.* surface, *b.* lateral view; after Tulasne.

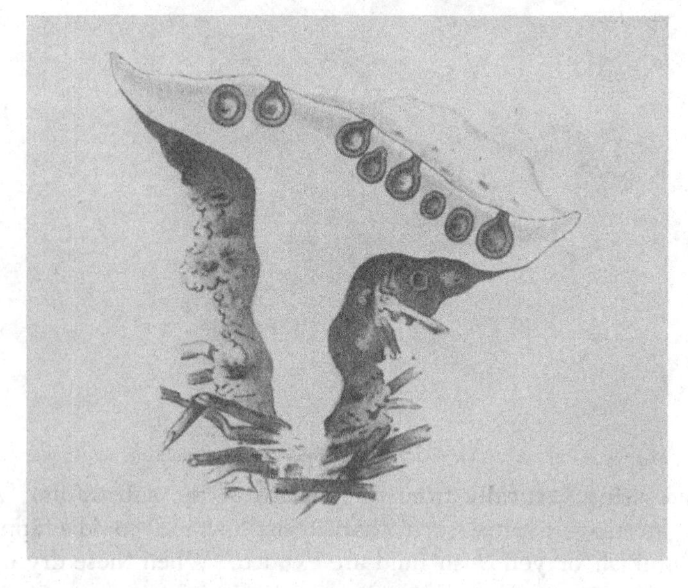

Fig. 126. *Poronia punctata* (L.) Fr.; stroma cut across; after Tulasne.

The perithecium is initiated by the development of a coil of large, deeply-staining cells forming the archicarp. It arises amongst the vegetative filaments of the stroma, forms a couple of loops and is continued towards the surface of the stroma as a slender multicellular trichogyne (fig. 127 *a*). At an early stage the coiled portion becomes surrounded by a knot of small, densely-staining hyphae; later the trichogyne disappears, degeneration progressing from the base to the apex; the investing filaments grow more actively on the side of the archicarp towards the surface of the stroma, so that the young perithecium becomes pear-shaped (fig. 127 *b*, *c*); further growth renders it hollow, and the upper part becomes lined with delicate periphyses (fig. 127 *d*). At the base of the developing perithecium is a group

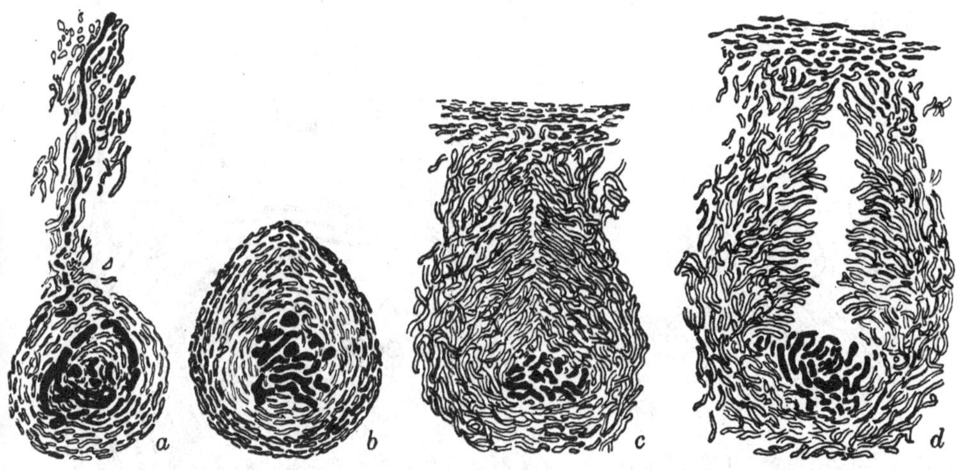


Fig. 127. *Poronia punctata* (L.) Fr.; *a.* archicarp, × 275; *b. c.* and *d.* young perithecia, × 205; after Dawson.

of stout, deeply-staining hyphae, from which the asci arise and which occupy the position of the coiled archicarp in earlier stages. Later the base and sides of the perithecium are covered by numbers of filiform, septate paraphyses, and amongst these the asci develop.

It seems pretty clear that the trichogyne now no longer functions; this is borne out by the fact that degeneration proceeds from its base upwards and not from its apex, as might have been expected if a male nucleus were travelling down. It is probable, though it has not actually been demonstrated, that the ascogenous hyphae are derived from the archicarp, but in view of the complete degeneration of this organ in *Gnomonia*, it is not safe to conclude without further evidence that it is still functional in *Poronia punctata*. The species deserves further investigation, especially from this point of view.

In both *Xylaria* and *Hypoxylon* the young stroma is covered by a tangle of conidiophores, from which small oval conidia are abstricted. In *Xylaria* these form a white coating, in marked contrast to the older black portions of the stroma, where the perithecia are maturing, and justify the name candle-snuff fungus, applied to some of the commoner species. If, in either genus, the stroma be sectioned during the conidial stage, nests of small hyphae, similar to those in *Poronia*, will be found, and are the first indications of the perithecia.

Sometimes a stouter hypha with larger nuclei, presumably an archicarp, is recognizable (fig. 128), but it has not been shown to function, and there is no evidence of normal sexuality. It is however not unlikely that some of these species, which are conveniently easy to microtome, will repay further investigation. At a later stage ascogenous hyphae are readily recognizable (fig. 129).

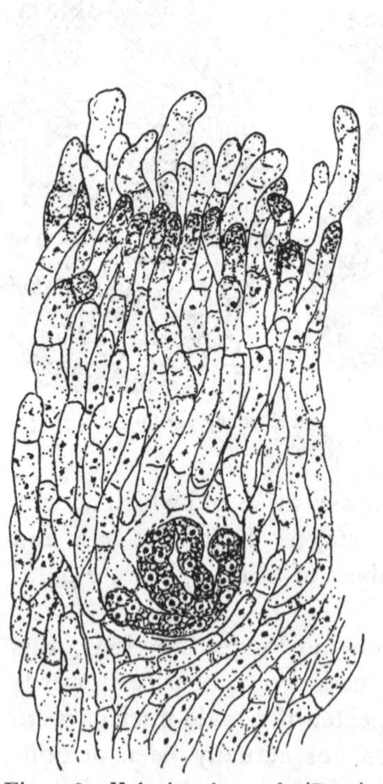

Fig. 128. *Xylaria polymorpha* (Pers.) Grev.; archicarp embedded in stroma, × 1000.

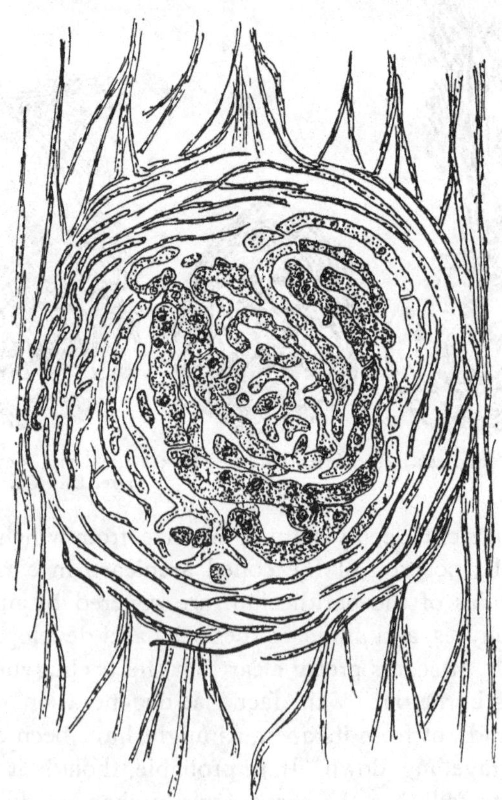

Fig. 129. *Xylaria polymorpha* (Pers.) Grev.; septate archicarp, × 1000.

## XYLARIACEAE: BIBLIOGRAPHY

1861–5 TULASNE, L. R. and C. Selecta Fungorum Carpologia; Imperial-typograph., Paris.
1900 DAWSON, M. On the Biology of *Poronia punctata* (L.). Ann. Bot. xiv, p. 245.

## LABOULBENIALES

The group Laboulbeniales includes some six hundred species arranged in over fifty genera. All are minute external parasites on insects, chiefly on members of the Coleoptera. They appear to do but little injury to the host, inducing at most a slight irritation but never causing death, indeed their own existence depends on that of the insect to which they are attached since, unlike many other fungi, their life ends with that of their host.

The Laboulbeniales are all of fairly simple structure (fig. 130) and show an underlying similarity of type. In all cases the vegetative part consists of a **receptacle**, usually two-celled, attached to the integument of the host by a blackened base or foot. From the receptacle grow out filamentous **appendages** on or among which the male organs are produced and, with a few exceptions, the receptacle of the same individual also gives rise to a female organ from which a perithecium liberating ascospores is eventually developed.

The plant is covered by a thin, homogeneous membrane which is exceedingly tough and impervious and is developed from the gelatinous coat of the spore; it efficiently protects the cells from desiccation.

Within this envelope the cell walls (except those of the receptive parts of the trichogyne and of the internal cells of the perithecium) are very thick and laminated. In certain cases, and especially in the genus *Laboulbenia*, they are traversed by fibrillae which arise from the innermost wall layer and are attached to the inner surface of the envelope. The cells are uninucleate (fig. 131) with rather dense, granular or reticulate cytoplasm and

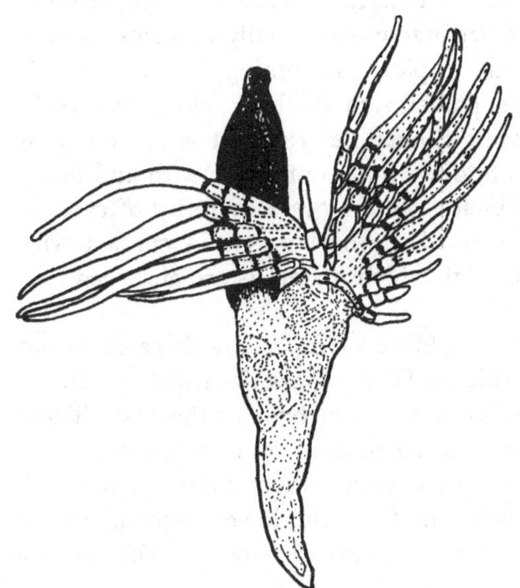

Fig. 130. *Laboulbenia triordinata* Thaxter; × 135; after Thaxter.

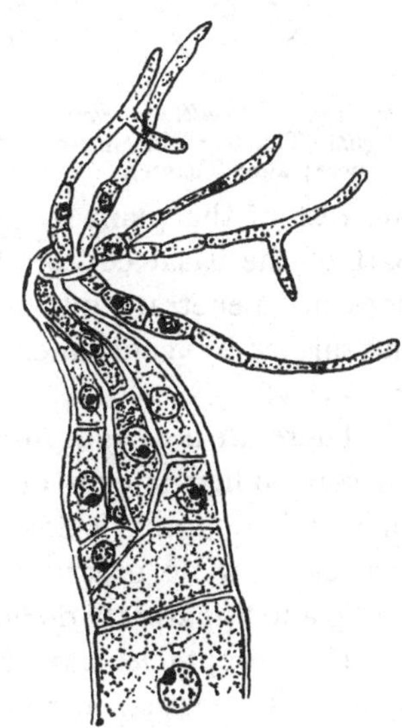

Fig. 131. *Laboulbenia chaetophora* young perithecium and trichogyne, × 360; after Faull.

contain oil globules. Between adjacent cells that have the same origin the protoplasm is continuous through broad pits. The cytoplasm on each side dips into the pit, forming a thick strand which, in *Laboulbenia* at least, appears to be intersected by the middle lamella (Faull). The latter in favourable cases is seen to be perforated by one or more fine pores through which complete continuity is established[1]. There is evidence that a stout strand of cytoplasm unites contiguous cells in the appendages.

The spores are remarkably uniform throughout the group, being invariably hyaline and fusiform or acicular in shape and almost invariably two-celled (fig. 132). The cells are commonly of unequal size, that nearest the apex of the ascus being the larger, and both are uninucleate. The spore is clothed in a gelatinous sheath especially well developed about the upper end which, when the spore is discharged from the perithecium, is destined to come in contact with the integument of the host. Here the gelatinous mass enables the spore to take up the oblique position in which germination occurs.

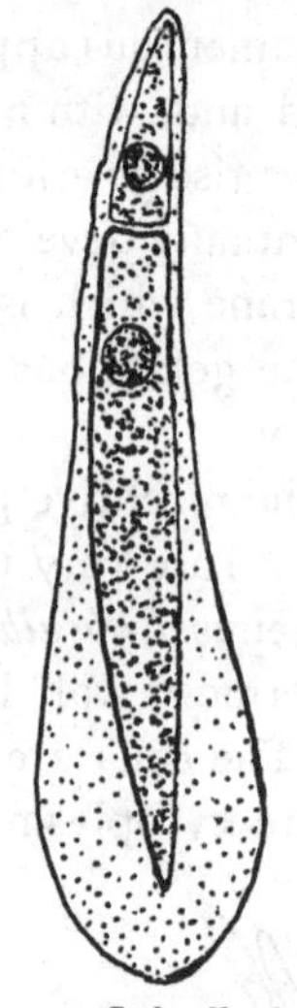

Fig. 132. *Laboulbenia elongata* Thaxter; bicellular spore; after Thaxter.

The lower extremity of the spore (its apex while in the ascus) forms the foot. As a rule the gelatinous envelope in this region becomes hard, opaque, black and more or less elastic and thus, while adhering firmly to the substratum, it may give the plant a certain freedom of movement. This elasticity is found especially in forms on submerged and rapidly swimming hosts where it allows the parasite to lie back along the body of the insect. Sometimes the foot is cut off from the rest of the plant by a wall, more often it is continuous with and forms part of the basal cell of the receptacle. In the great majority of cases it does not penetrate into the substance of the host but is in contact with the surface by a thin membrane through which materials are absorbed into its cavity.

There are a certain number of forms, however, especially those occurring on soft-bodied insects or on the soft parts of others, in which a definite rhizoidal apparatus is developed and penetrates the body of the host. These species show no greater vegetative luxuriance than other members of the group and apparently do not benefit by their more elaborate absorptive organ.

The receptacle, like the foot, develops from the lower segment of the spore. It consists, in the simplest cases, of two superposed cells and (in

[1] A similar type of pitting has been described by Kienitz-Gerloff for the Red Algae ("Neue Studien über Plasmodesmen," *Ber. d. deut. bot. Gesel.* xx, 1902.)

monoecious forms) bears the appendages in a terminal position and the perithecium laterally (fig. 136).

More rarely the receptacle consists of a larger number of cells variously arranged and reaching a considerable complexity in such forms as *Zodiomyces vorticellarius* (fig. 133).

One or more appendages are borne on the receptacle. These are more or less filamentous and often elaborately branched. They bear the male organs and serve also for the protection of the delicate trichogyne and perhaps facilitate fertilization by holding a drop of water around the organs concerned. The primary appendage is developed from the upper segment of the germinating spore and is terminal; the later formed secondary appendages, when present, are outgrowths from the cells of the receptacle.

The male element is a non-motile cell which as early as 1896 was homologized by Thaxter with the spermatium of the Red Algae. The latter organ has now been shown to be an antheridium[1] in which the nuclear divisions are reduced to one, or have altogether disappeared; it is liberated entire from the male plant and carried passively to the female organ. It seems very probable that in the simplest cases, where they are produced externally at the tips of more or less specialized branches (fig. 134),

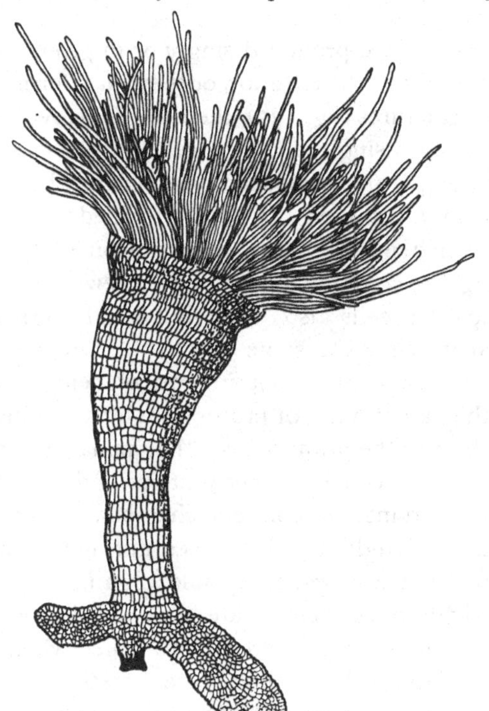

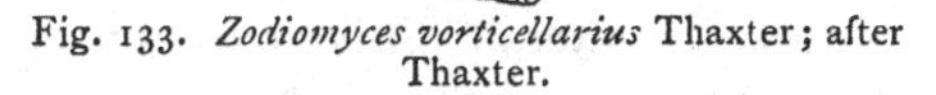

Fig. 133. *Zodiomyces vorticellarius* Thaxter; after Thaxter.

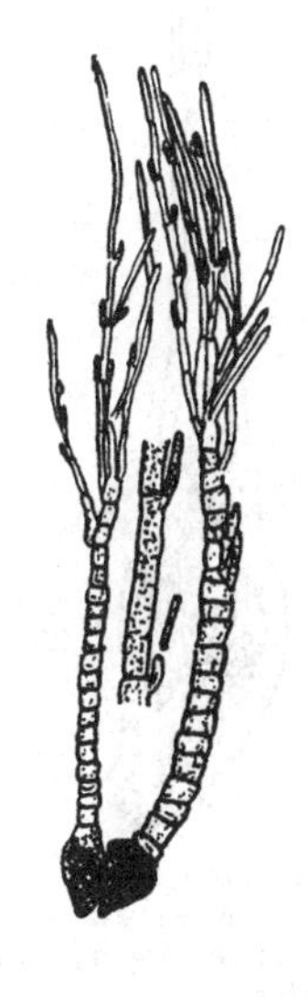

Fig. 134. *Ceratomyces rostratus* Thaxter; exogenous spermatia; after Thaxter.

[1] Wolfe, *Ann. Bot.* xviii, 1904. Yamanouchi, *Bot. Gaz.* lxii, 1906.

the "antherozoids" or spermatia of the Laboulbeniales have the same significance as those of other fungi. They fall off when mature and the cells from which they were formed may give rise to others in the same position.

In *Coreomyces* instead of a segment of the branch being detached to form the spermatium, a portion of its contents is extruded. This arrangement leads to the more specialized endogenous organ which is found in *Stigmatomyces* (fig. 136) and in many other forms. Here the naked mass of protoplasm cut off from the parent cell may be regarded as the homologue of the spermatium, or the parent cell may be recognized as an antheridium and the detached segments as non-motile spermatozoids. They function, in any case, as sperms or male elements. They are detached from the contents of a flask-shaped cell and are extruded through an elongated neck opening at maturity to the exterior. Between the neck and the venter a diaphragm of cellulose is deposited and is perforated by a narrow opening so that the sperms are nipped off as they pass into the neck. They are uninucleate, the nucleus of the parent cell undergoing successive divisions so that a series of sperms are produced. The parent cell has been termed an antheridium but if the spermatium is antheridial in character the name cannot appropriately be used for the cell in which it is borne, though the term spermogonium is applicable.

These sperm-forming organs may be produced singly or in groups, each with its neck opening independently to the exterior, or they may form compound structures (fig. 135), the necks of several cells penetrating a single adjacent cell into the cavity of which the sperms are discharged and from which they escape by a common duct, the so-called secondary neck, which may be a mere extension of the cell forming the common cavity or, in a few cases, may involve other cells also. The individual sperms are formed in much the same way in the compound as in the simple organs, but instead of being cut off from the parent mass of protoplasm by a diaphragm at the base of the primary neck they remain attached till the neck widens abruptly at its end, and they are extruded into the common chamber. Hundreds, or even thousands, may be formed during the period of activity of a single compound organ. The exogenously produced sperms are always walled whereas those formed endogenously are naked when first set free; later a thin wall may be secreted.

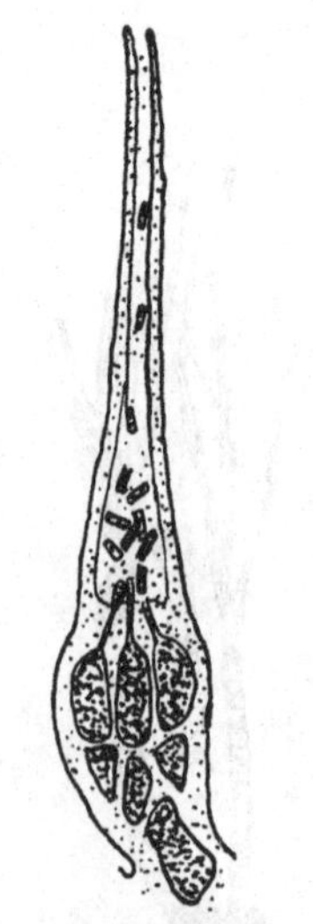

Fig. 135. *Dimeromyces Africanus* Thaxter; compound spermatial organ; after Thaxter.

The female organs are formed from the basal cell of the spore and are thus necessarily lateral. This condition is often obscured in the mature plant where the developing perithecium may push the appendages aside and take up an apparently terminal position. The development is very uniform, and has been described by Thaxter in some detail for *Stigmatomyces Baeri*. Here, the upper cell of the receptacle divides into two; the lower of these remains as part of the receptacle, and the upper grows out (fig. 136*a*) to form the female organ and ultimately

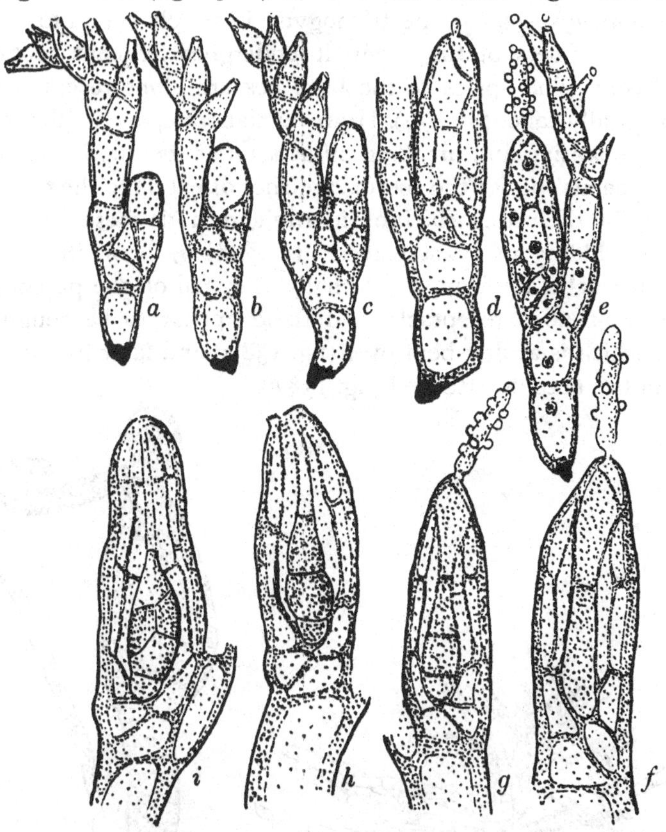


Fig. 136. *Stigmatomyces Baeri* Peyritsch; development of the perithecium; *a.* shows the two-celled receptacle, a single appendage bearing five simple, endogenous spermatial organs, and the beginning of the perithecium; *b.*—*i.* indicate successive stages in the development of the perithecium; the trichogyne first appears in *d.*; in *e.* spermatia are being shot out and some are attached to the trichogyne; in *i.* two of the four ascogenous cells are shown, with the superior sterile cell above them, and the primary and secondary inferior sterile cells below; after Thaxter.

the perithecium. It divides transversely; the upper of its daughter cells gives rise to the female organ, the lower divides several times (fig. 136*b*), and ultimately forms the double wall of the perithecium, a function fulfilled by a complex of neighbouring hyphae in Ascomycetes with a richer vegetative development.

The upper cell, the initial of the female organ, divides, separating the female cell below (fig. 136*c*) and a cell above, which divides once more to form the terminal trichogyne and the subjacent trichophoric cell (fig. 136*d*).

All the cells are uninucleate. The female cell is called by Thaxter a carpogonium or carpogonic cell in conformity with the term used for the Red Algae, but it obviously corresponds to the cell in which fertilization is now known to occur in other Ascomycetes and will therefore here be termed the oogonium.

In *Stigmatomyces Baeri* the trichogyne is simple (fig. 136 *d*, *e*) but in many other members of the group it undergoes frequent septation and branches freely. The apices of the branches are alone receptive and may straight or spirally coiled (fig. 137). However elaborate, the trichogyne quickly disappears, collapsing and breaking off as soon as its function is fulfilled.

In endogenous species the sperms are shot direct on to the trichogyne or carried to it by the water which ordinarily surrounds these filaments when the hosts are hiding in moist places. In *Zodiomyces* on the other hand, where the spermatia are formed externally, they fall off the parent branches on to the cup-shaped receptacle, and there appear to be sought by the trichogyne which is at first bent over (fig. 138 *a*) and later lifts itself after a spermatium has become attached (fig. 138 *b*).

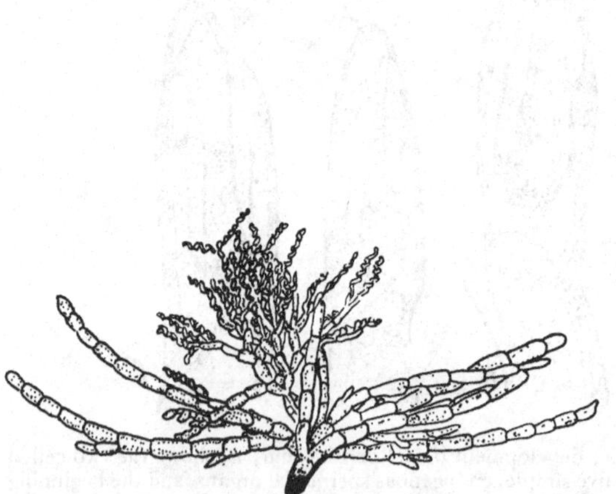

Fig. 137. *Compsomyces verticillatus* Thaxter; after Thaxter.

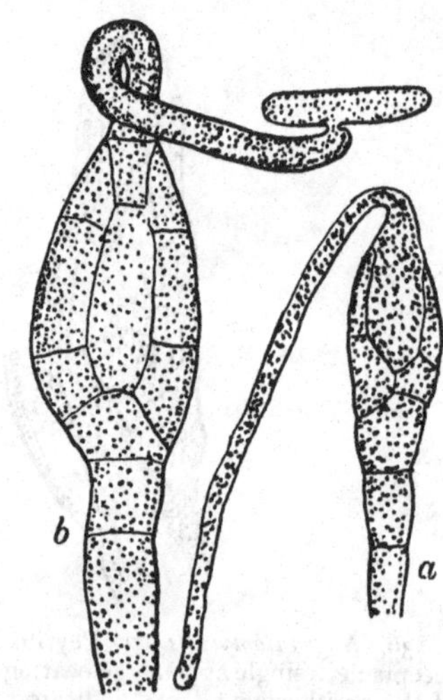


Fig. 138. *Zodiomyces vorticellarius* Thaxter; trichogyne *a*. before and *b*. after attachment of spermatium; after Thaxter.

In any case numerous male cells reach the trichogyne and, though the actual process of fertilization has not yet been seen, it appears likely that it is accomplished.

Afterwards the oogonium divides into three superposed cells, the sterile inferior cell, the sterile superior cell and a fertile cell lying between the two (fig. 136 *g*, *h*). This middle cell cuts off a secondary sterile cell below (fig. 136 *i*) which like the other sterile cells is eventually destroyed. It then divides longitudinally into four "ascogenic" cells, two of which are shown in fig. 136 *i*, and from these asci bud out, arising in a more or less distinctly double row (fig. 139 *a*). Some variation occurs in different species in the later divisions and in the number of ascogenic cells. In *Polyascomyces* (fig. 140) more than thirty are present, covering a basal area from which numerous asci bud upwards, so that the condition approximates that in other Ascomycetes. Faull describes the ascogenic cells as binucleate, each containing two

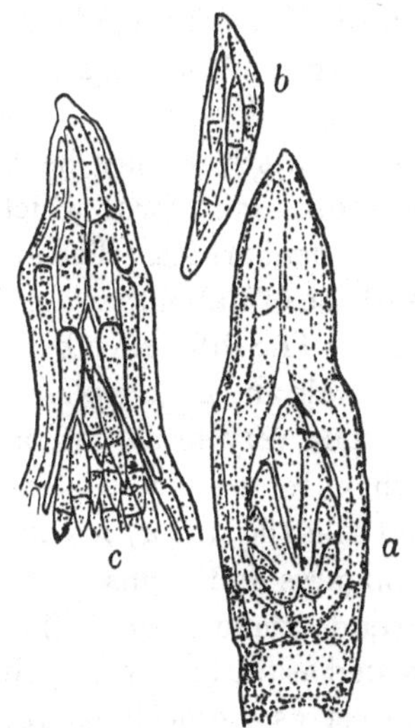


Fig. 139. *Stigmatomyces Baeri* Peyritsch; *a*. young asci; *b*. ascus containing four spores; *c*. mass of spores in perithecium; after Thaxter.

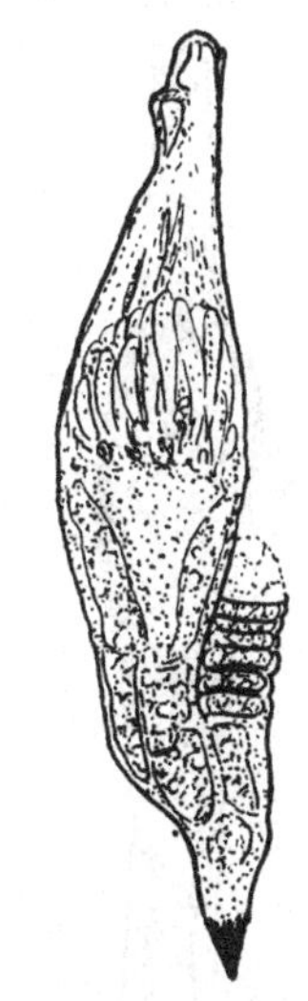

Fig. 140. *Polyascomyces Trichophyae* Thaxter; after Thaxter.

nuclei which undergo conjugate division whenever an ascus is formed. As a result the young ascus is binucleate and nuclear fusion followed by three divisions takes place in the usual way. As a rule four only of the eight nuclei function; the spores are produced in a manner quite characteristic of the Ascomycetes generally. In the ascus they are usually disposed more or less definitely in pairs and the members of a pair are discharged together from the perithecium and germinate side by side.

In monoecious species one member of a spore pair may frequently produce a smaller and weaker individual than the other, while in *Laboulbenia inflata* the atrophy of one at an early stage of development is a regular phenomenon. In *Stigmatomyces Sarcophagae* the smaller individual is unisexual, producing only male cells, while the larger is hermaphrodite (fig. 141).

In dioecious species the paired spores are of rather different sizes. The smaller spore gives rise to a male plant, the larger to a female, so that by their association at a point of contact with the host a condition essential for the perpetuation of such species is secured. The cytological changes by which this segregation of sex is brought about between the members of a pair should be of great interest and demand investigation.

There is an obvious suggestion in these phenomena of a transition between the monoecious and dioecious condition but it is not clear in which direction the series should be read. It might be inferred that the male plant had become atrophied after the female had acquired spermatial organs, or on the other hand that, as in many other groups of plants, a hermaphrodite condition was primitive and segregation a later development.

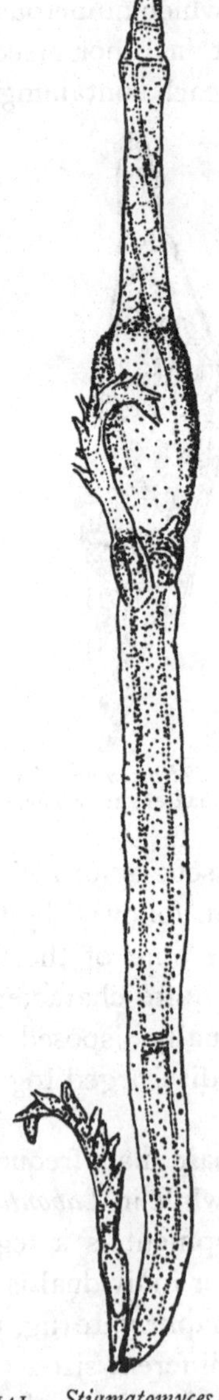

Fig. 141. *Stigmatomyces Sarcophagae* Thaxter; male and hermaphrodite individuals, × 260; after Thaxter.

*Amorphomyces Falagriae* may be taken as an example of a dioecious form which shows also several other peculiarities. The spores are unique amongst those of Laboulbeniales in being aseptate (fig. 142). The difference between the spores producing male and female plants is slight at first but becomes very apparent on germination. In each case the spore divides into three superposed cells (fig. 143 *a*), in the male the terminal cell elongates and forms a single male organ liberating endogenous sperms. The second cell may be regarded as the basal cell of this structure and the lowest as a unicellular receptacle, or they may be held to constitute together a two-celled receptacle. There are no appendages.

In the female the lowest cell, which may become partly divided, forms the receptacle, the next above gives rise to the perithecial wall and the terminal cell to the female organ proper. The perithecium and its contents are therefore here terminal, a state of affairs not met with elsewhere in the group. The terminal cell divides in the usual way to form an oogonium, a trichophoric cell and a trichogyne; the latter is short and branching (fig. 143 *b*). The development of the perithecium (fig. 144) seems fairly typical and the asci apparently contain four spores.

In 1912, Faull published an account of the cytology of two species of *Laboulbenia*, *L. chaetophora*, and *L. Gyrinidarum*. Both occur on the same host, and could not be distinguished in the young stages. Both are parthenogenetic, no male cells being formed.

A trichogyne, trichophoric cell and oogonium are formed in the usual way (fig. 131). According to Faull nuclear division takes place both in the oogonium, and in the trichophoric cell, and the partition between these two breaks down so that a long cell containing a row of four nuclei is formed (fig. 145 *a*).

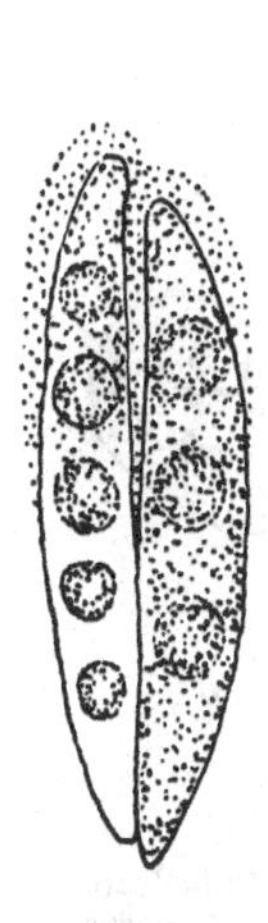

Fig. 142. *Amorphomyces Falagriae* Thaxter; paired spores; after Thaxter.

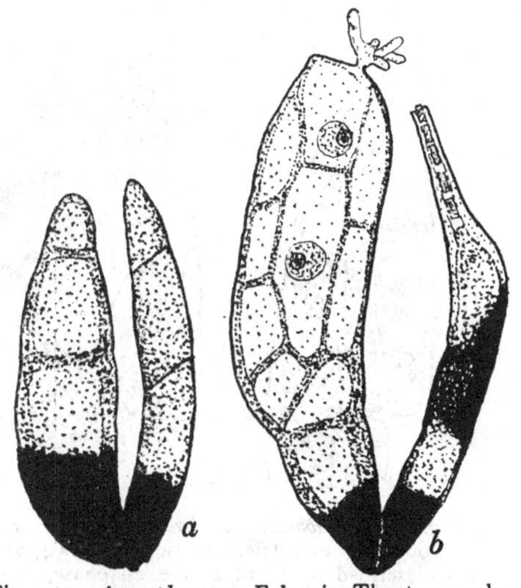


Fig. 143. *Amorphomyces Falagriae* Thaxter; male and female individuals; *a*. young, *b*. mature; after Thaxter.

Walls cut off the upper and the lower nucleus, and a central binucleate cell is left, the lower nucleus of which is presumably a daughter of the oogonial and the upper of the trichophoric nucleus. These divide simultaneously and a binucleate inferior sterile cell is separated from the binucleate fertile cell. This in turn divides to form the ascogenic cells, from which the asci are to develop, and these and the asci which they produce are therefore binucleate. The two nuclei in the ascus fuse and their union is regarded by Faull as the only nuclear fusion which occurs in this very curious life history. Meiosis then takes place, followed by the third division. The upper daughter nuclei of this division degenerate and around the lower nuclei spores are organized. In each spore the nucleus divides once and a transverse septum is formed. Faull describes four chromosomes (fig. 145 *c*) at every stage but figures an apparently larger number in the first division in the ascus where the structures represented are evidently gemini (fig. 145 *b*).

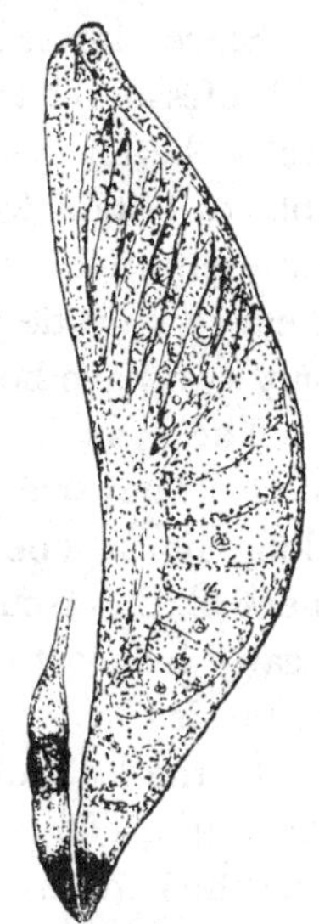

Fig. 144. *Amorphomyces Falagriae* Thaxter; male and female individuals, the latter with perithecium containing spores; after Thaxter.

The Laboulbeniales are subdivided by Thaxter according to the method of formation of the male cells, whether exogenous or endogenous, and in the latter case whether produced in simple or compound organs. In this way three families, Peyritschielaceae (compound endogenous), Laboulbeniaceae (simple endogenous) and Ceratomycetaceae (exogenous) are distinguished.

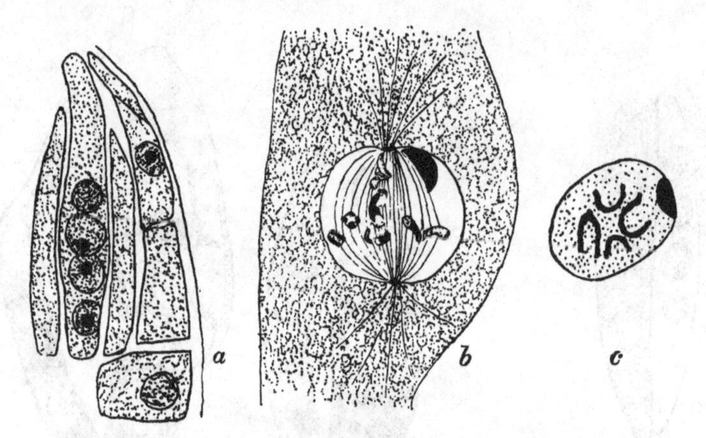


Fig. 145. *Laboulbenia chaetophora* (?). *a*. cell formed by binucleate oogonial and trichophoric cells, × 430; *b*. first division in ascus described by Faull as the anaphase, × 1510; *c*. nuclear division in spore, showing four chromosomes, × 2800; after Faull.

Since almost all our knowledge of the group is due to the brilliant work of Professor Thaxter of Harvard it follows that the North American species are far better known than those of other localities. Such material as he was able to obtain from warmer regions proved, however, exceedingly rich in representatives of the group, and it is probable that further study will show them to be widely distributed. The known European species are few, and only two have been identified in Great Britain.

The systematic relations of the Laboulbeniales are not easy to determine. They are pretty evidently monophyletic, and are highly specialized along lines dependent on their peculiar habitat. The form and development of the ascus is typical of the Ascomycetes, and the Laboulbeniales clearly belong to that group, though it is difficult to indicate their affinities within it.

In the Laboulbeniales the young female branch consists of four parts, the initial cell of the perithecium wall, the oogonium (carpogenic cell of Thaxter), the trichophoric cell and the trichogyne. These correspond pretty clearly with the archicarp of other groups. The initial cell of the perithecium wall constitutes a stalk-cell from which the enveloping hyphae develop. The trichophoric cell should probably be included as part of the trichogyne, which in this sense always consists of at least two cells. After fertilization the oogonium divides to form a row of cells, which are, from below upwards, the inferior sterile cell, the secondary inferior sterile cell, the mother-cell[1] of the ascogenic cells and the superior sterile cell. The subterminal of these alone gives rise to asci; this it does by dividing longitudinally into a definite number of ascogenic cells from which the asci are budded out.

A multicellular trichogyne is not uncommon among Ascomycetes, and the division of the oogonium after fertilization into a row of cells is a well-known phenomenon occurring among Plectascales, Erysiphales, and Discomycetes. In most cases several cells of the septate oogonium give rise to ascogenous hyphae, but in the Erysiphales only the subterminal cell of the row does so. In *Erysiphe* this cell, which always contains at least two nuclei, gives rise to several asci and differs from the subterminal cell of the Laboulbeniales chiefly in the fact that the asci are produced from short outgrowths instead of, as in *Stigmatomyces* and its allies, from daughter cells formed by longitudinal division. In this character then, the Laboulbeniales would appear most nearly to approach the Erysiphaceae, which they also resemble in the formation of the sheath from the stalk cell of the oogonium, but they differ from them in possessing a trichogyne, an organ not known in that group, where the antheridium comes into direct contact with the oogonium.

The Laboulbeniales and Erysiphaceae have also in common the uninucleate character of their vegetative cells.

In the structure of their ascocarp, opening as it does by a narrow aperture, the Laboulbeniales approach most closely to the other Pyrenomycetes.

The male element in the Laboulbeniales is a non-motile, uninucleate structure which may be budded off externally from the parent cell, or extruded from it as a naked mass of protoplasm.

Amongst the Fungi spermatia occur in the Pyrenomycetes and Lichens and also characterize the Uredinales, but in all these cases they are budded off externally as walled structures. The spermatium both in the Red Algae and in the Fungi has been homologized with a reduced antheridium, and, as has already been pointed out, the exogenous male element no doubt bears the same significance among the Laboulbeniales. We have no satisfactory indication as to the relative primitiveness of the endogenous and the exogenous condition, but it may be noted that exogenous forms only are known among Fungi other than Laboulbeniales. The endogenous organ may be derived from the branch which cuts off exogenous, walled spermatia, or it may represent quite a different response to the need for non-motile male elements, the parent cell being the homologue of the antheridium and the fertilizing element that of a spermatozoid.

These various characters, approximating to those of sometimes one,

[1] This cell is described by Thaxter as the ascogonium. The word has acquired a somewhat different sense in other Ascomycetes.

sometimes another group of Ascomycetes, seem on the whole to indicate no very close relationship, but suggest rather that the Laboulbeniales were derived from an ancestral form, already definitely ascomycetous but not otherwise highly specialized, and that they have undergone elaborate and characteristic modifications after branching off from the main line. Their nearest affinities are with the Erysiphales and Pyrenomycetes.

The resemblances between the Laboulbeniales and the Red Algae have been regarded as significant in connection with the hypothetical relationship of the higher Fungi to that group.

## LABOULBENIALES: BIBLIOGRAPHY

1896 THAXTER, R. Contribution towards a Monograph of the Laboulbeniaceae, Pt. I. Mem. Am. Acad. Arts and Sciences, xii, p. 195.

1908 BIFFEN, R. H. First Record of Two Species of Laboulbeniaceae for Britain. Trans. Brit. Myc. Soc. iii, p. 83.

1908 THAXTER, R. Contribution towards a Monograph of the Laboulbeniaceae, Pt. II. Mem. Am. Acad. Arts and Sciences, xiii, p. 219.

1911 FAULL, J. H. The Cytology of the Laboulbeniales. Ann. Bot. xxv, p. 649.

1912 FAULL, J. H. The Cytology of *Laboulbenia chaetophora* and *L. Gyrinidarum*. Ann. Bot. xxvi, p. 325.

# CHAPTER VI

## BASIDIOMYCETES

THE Basidiomycetes include over 13,000 species possessing a well-developed mycelium, which, among the higher forms, builds up an elaborate fruit-body such as may be observed in the toadstools, bracket fungi and puff balls.

They are characterized by the fact that their principal spores, the **basidiospores**, are borne externally on the mother-cell or **basidium**. The young basidium contains two nuclei; these fuse, and the fusion nucleus divides twice, providing the nuclei of the four spores; each spore is formed at the end of a stalk or **sterigma** through which the nucleus passes to enter the developing spore. The two divisions in the basidium constitute a meiotic phase. In the Autobasidiomycetes the basidium is without septa, and the spores, except where some fail to develop, are regularly four in number for each basidium. In the Protobasidiomycetes the basidium is divided into four cells, each of which gives rise to a single spore; the walls are transverse in the Uredinales and Auriculariales, longitudinal or oblique in the Tremellales. In the Hemibasidiomycetes (Ustilaginales) septa may or may not be present in the basidium, but the fusion nucleus divides more than twice, and more than four spores are produced.

In *Puccinia*, *Phragmidium* and other Uredinales, and in *Sirobasidium* and its allies, the basidia are developed in chains, in other cases they are borne singly. In the Ustilaginales and in the majority of the Uredinales the nucleus and cytoplasm of the basidium are at first enclosed in a thick wall forming the **brand-spore** or **teleutospore cell**, which becomes detached, forming an additional means for the distribution of the plant; later the contents are extruded as a thin-walled **promycelium** on which the basidiospores are produced. In other Basidiomycetes the basidia are thin-walled throughout their development and produce spores while still attached to the mycelium.

The basidiospore is unicellular, round or oval in shape, usually with a smooth, rather thin wall. Echinulate or warted spores occur in a few species, and in many families, especially among gill-bearing fungi, dark or bright-coloured spores are common.

In a considerable number of genera accessory spores are also produced.

Except for the production of their characteristic spores externally on basidia, the Ustilaginales and Uredinales differ in almost every particular from the majority of the Basidiomycetes; they are obligate parasites with a delicate mycelium ramifying in the tissues of the host and they lack the elaborate stroma characteristic of the Autobasidiomycetes. It has therefore seemed advisable to deal with them as distinct groups, separating them for purposes of description from the Basidiomycetes proper.

# CHAPTER VII

## HEMIBASIDIOMYCETES

### USTILAGINALES

THE Ustilaginales, Brand fungi, Smuts or Bunts, constitute a group of some 400 obligate parasites on the higher plants, giving rise in the tissues of the host to characteristic, usually dark-coloured resting-spores, the **brand-spores**, teleutospores or chlamydospores. These are developed in considerable quantities, either singly, in pairs, or in clusters known as **spore-balls**, and when ripe break through the host tissue, forming a pustule or **sorus**. No distortion of the host is caused during the period of vegetative growth, but in preparation for the formation of spores very considerable hypertrophy may be induced.

*Ustilago Treubii* on the stem of *Polygonum chinense* in Java causes the formation of elaborate galls (fig. 146) provided with vascular tissue and growing by means of a cambium; *Ustilago Maydis* produces whitish swellings and blisters, often as large as a fist, on the stem, leaves, roots, and especially the flowers of *Zea Mays*; and *Urocystis Violae* deforms the stems and leaves of various species of *Viola*.

Fig. 146. *Ustilago Treubii* Solms; stem of *Polygonum* with "fruit gall," nat. size; after Solms Laubach.

Several other smuts develop their spores in the ovary of the host plant, or infect the stamens, filling the anthers with spores and benefiting by the means of distribution provided for the pollen. *Ustilago antherarum*[1] even induces development in the staminal rudiments of the normally pistillate flowers of *Lychnis dioica*. The stamens formed undergo dehiscence as usual and differ from those of the male flowers only in the presence of fungal spores instead of pollen in their anthers.

In all these cases and in most of the Ustilaginales spore-formation is strictly localized, but in the genus *Entyloma* and its allies spores may be formed at almost any point.

[1] *Ustilago antherarum* (DC.) Fr. = *Ustilago violacea* (Pers.) Fuck.

The mature brand-spore is uninucleate, and is surrounded by a delicate endospore and by an epispore which may be smooth or variously sculptured and usually contains pigment, giving the spore a black, brown, or violet colour.

On germination the spore gives rise to a short tube, the promycelium or basidium (fig. 147), into which its contents pass, the nucleus undergoing at least two divisions; the basidium in turn produces a number of uninucleate sporidia or basidiospores. The basidium may be unicellular, giving rise to a bunch of basidiospores at its apex (*Tilletia* (fig. 148 *a*)), or multicellular, usually four-celled, producing one or more basidiospores from each cell (*Ustilago* (fig. 147 *e*)). The nucleus of the parent cell does not travel into the basidiospore but divides, sending one daughter nucleus into the spore, while the other, remaining in the basidial cell, may undergo further divisions so that nuclei are provided for a number of spores.

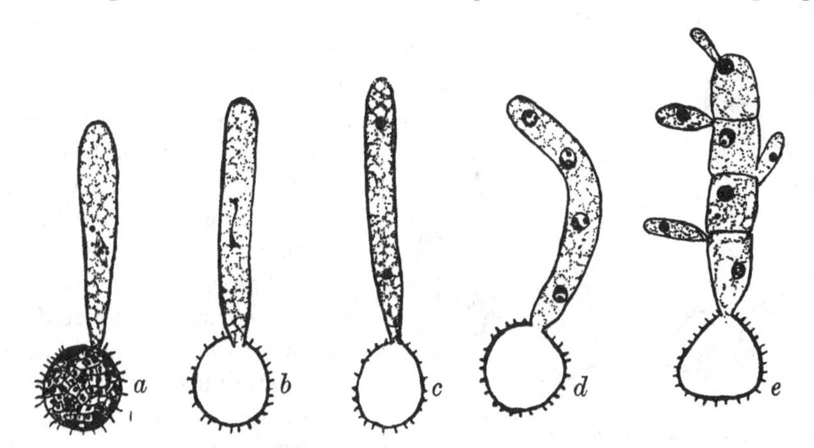


Fig. 147. *Ustilago Scabiosae* Sow.; development of basidium; after Harper.

Under suitable conditions the basidiospores are cut off in considerable numbers. They may further multiply by budding, giving rise to conidia, or a delicate mycelium may be formed from which conidia are abstricted (*Tilletia*). A supply of conidia is produced by these means in dung decoction and other nutrient solutions, and no doubt in the damp, manured soil of the

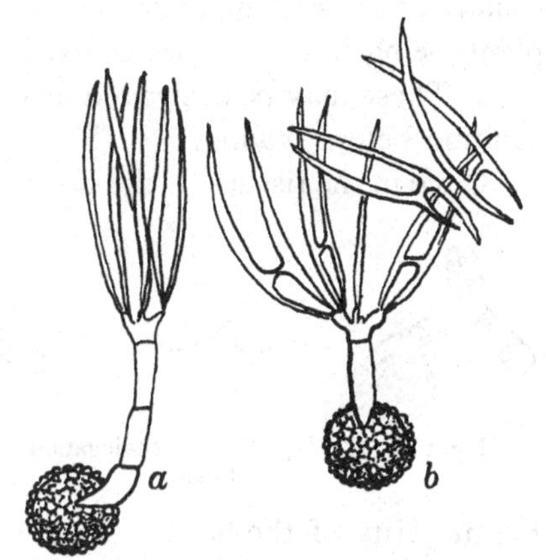


Fig. 148. *Tilletia Tritici* (Bjerk) Wint.; *a.* basidium thirty hours after germination of brand-spore; *b.* after conjugation of basidiospores; × 300; after Plowright.

fields. As a rule the conidia are of the same oblong form as the basidiospores, but, in the genus *Tilletia* and some of its allies, they may be stout or sickle-shaped, whereas the basidiospores are long and narrow.

In *Entyloma* the brand-spores are capable of germination on the tissues of the host leaf, where they give rise to hyphae which penetrate through the stomata and form basidia from which basidiospores are produced.

During their development the cells of the basidium, the basidiospores, or the conidia budded out from them, may become united in pairs (figs. 148 *b*, 149), by means of a tube of variable length put out by one or both participants and recalling somewhat the conjugation tube of the Zygnemaceae. The growth of these tubes is very accurately directed and appears to depend on a chemotropic stimulus.

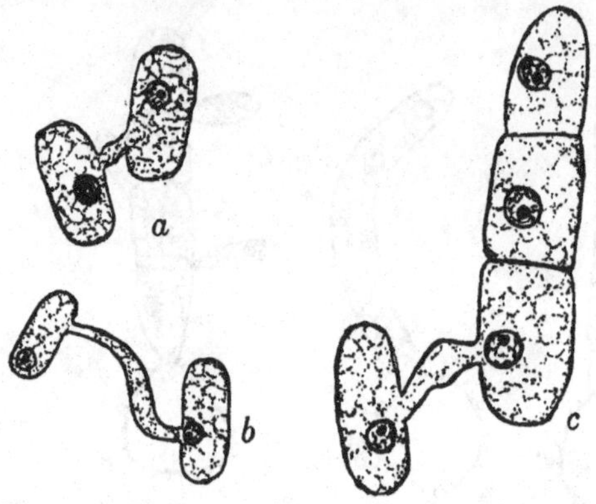


Fig. 149. *Ustilago antherarum* Fr.; *a.* and *b.* conjugating basidiospores; *c.* conjugation between a cell of the basidium and a basidiospore; after Harper.

In the majority of cases the nucleus of one of the associated cells passes down the tube into the other, but does not fuse with its nucleus (fig. 150). Later both nuclei divide, and a mycelium of binucleate cells is produced. It is on this mycelium that the infection of the host depends; it penetrates the tissues usually of the seedling, but sometimes of the developing parts of the mature plant, being in most cases derived from spores which adhered to the seed coat. These may be destroyed by dipping the seed into hot water or formalin solution before sowing.

Once in the tissues of the host the binucleate mycelium penetrates in all directions, ramifying between the cells of the host and sending haustoria into them. The internodes of the stem are traversed by long, unbranched hyphae, but in the nodes branching is frequent, and here also the majority of the haustoria are to be found. Where the host is perennial the mycelium perennates in it, and, if the host dies down during the winter, remains alive but quiescent in the upper part of the root stock till the growth of new shoots in spring gives it a fresh opportunity of development.

Fig. 150. *Ustilago Hordei*; conjugation; after Lutman.

Conidia have been recorded on the parasitic mycelium of *Tuburcinia* and *Entyloma* but are not of common occurrence at this stage.

In the regions where the formation of brand-spores is to take place, the mycelium becomes richly branched and often swollen and gelatinous. In *Ustilago* and *Sphacelotheca* the sporogenous hyphae are divided into a number of short segments in each of which the contents form a spore surrounded by an independent membrane. The spores are enclosed at first within the gelatinous parent walls, but later these disappear so that the whole mycelium is transformed into a pulverulent mass of spores. In *Tilletia* and *Entyloma* the sporogenous cells are budded off laterally from the mycelium.

In *Tuburcinia* a number of richly septate hyphal branches become interwoven, forming a knot or **spore-ball** in which spores to the number of 50 or 100 are developed from within outwards. In *Urocystis* the spore-ball is small and the outer cells remain pale or colourless and do not function as spores though they resemble them in form (fig. 151). In *Doassansia* the outer sterile cells are wedge-shaped, and in *Sorosporium* they form a gelatinous investment in which their individual boundaries are no longer recognizable.

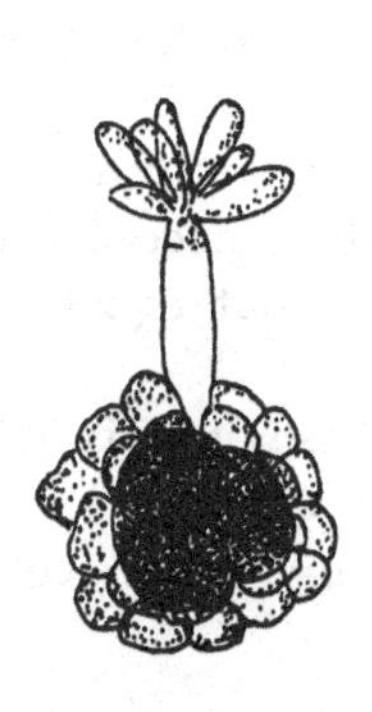

Fig. 151. *Urocystis Fischeri*; spore-ball, one spore germinating, × 500; after Plowright.

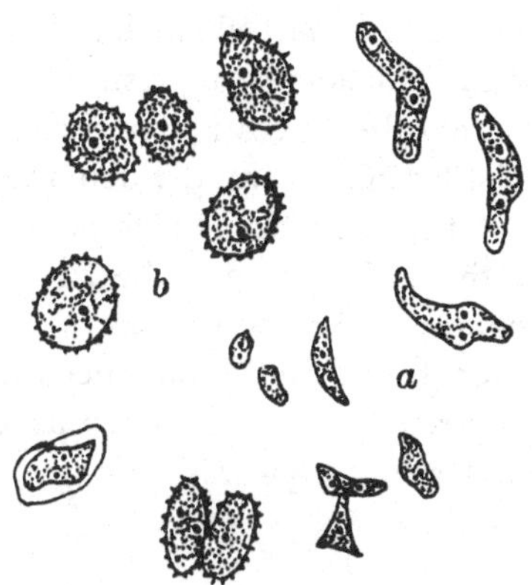


Fig. 152. *Ustilago Carbo*; *a*. young, binucleate brand-spores; *b*. older spores after nuclear fusion; after Rawitscher.

The young spore, like the cells of the mycelium from which it is derived, contains two nuclei (fig. 152 *a*). These undergo fusion, so that the mature spore is uninucleate (fig. 152 *b*). The pairing of the nuclei, which begins with the association of the basidiospores (or their conidia), is thus completed in the brand-spore.

The minute investigation of the group may be said to have begun in 1807 when Prévost recorded the germination of the spores of *Tilletia Tritici*. His work was continued by Berkeley, Tulasne, de Bary, the latter's pupil Fischer von Waldheim, and Brefeld.

Many of these early investigators observed the union of the sporidia in pairs, and, the nuclei not being identified, there was some question as to whether the process was to be regarded as sexual (de Bary), or as a merely vegetative phenomenon (Brefeld), like the formation of H-pieces and clamp-connections.

In 1894 Dangeard described the fusion of nuclei in the young brand-spore, but it was not till several years later that this was correlated with the union of the sporidia and the nuclear life-history made clear. The salient points of this history are (1) nuclear association, (2) nuclear fusion, (3) germination of the brand-spore, and formation of the basidium. At this stage the fusion nucleus divides twice or oftener, and uninucleate cells are formed. This sequence of events indicates that the basidium and its spores are the starting point of a brief haploid phase, which gives way to the diploid generation when conjugation takes place.

The life-history of the Ustilaginales would appear to be reduced rather than primitive, the conjugation of the spores replacing some ordinary sexual process; but the present state of our knowledge scarcely permits speculation as to what the earlier alternation of generations may have been. *Tuburcinia primulicola*, which has two parasitic phases, respectively uninucleate and binucleate, suggests closer comparison with the Uredinales than with any other investigated form.

The Ustilaginales are divided into two groups, distinguished by the character of the basidium, which is septate in the Ustilaginaceae and continuous in the Tilletiaceae. The two families are of about equal size, including together over 400 species.

The difficulty at first experienced in classifying these fungi is indicated by the occurrence of such names as *Uredo*, *Caeoma*, *Erysiphe*, *Ascomyces* (= *Exoascus*), and *Lycoperdon* among the older synonymy of the species.

### *Ustilaginaceae*

*Ustilago*, with nearly 200 species, is the most important genus of the Ustilaginaceae. It is cosmopolitan, occurring on all sorts of host plants, and is characterized by the fact that its brand-spores are produced singly.

*Ustilago Carbo* infects species of *Avena*, *Triticum* and *Hordeum*, the forms on the different hosts being biologically distinct. Rawitscher observed that the spores germinate readily in dilute nutritive solutions, forming a three or four celled basidium from which basidiospores may be abstricted in the usual way. More commonly, however, the basidia develop without spore-formation into branched mycelia, between the cells of which conjugation may take place. Union is accomplished between neighbouring cells of the same filament by means of short outgrowths, which meet and fuse as in the formation of clamp-connections (fig. 153 *a*), or between unrelated cells through a conjugation tube (fig. 153 *b*). Where basidiospores are formed they conjugate in a similar manner. In every case the nucleus of one of the paired cells passes over into the other, and the two nuclei lie close together, though without fusion. The mycelium throughout the development of the host plant consists of binucleate cells and breaks up in spore-formation into binucleate segments (fig. 152 *a*). Each young spore has thus two nuclei which fuse during development so that the mature brand-spore is uninucleate.

In *Ustilago Avenae*, *U. Hordei* and *U. Tritici*, sub-species of *U. Carbo*, Lutman observed that on conjugation some of the cytoplasm of one of the cells passed over with the nucleus, the empty cell becoming shrivelled (fig. 154). In *U. Avenae* a long fusion tube is frequently formed and both

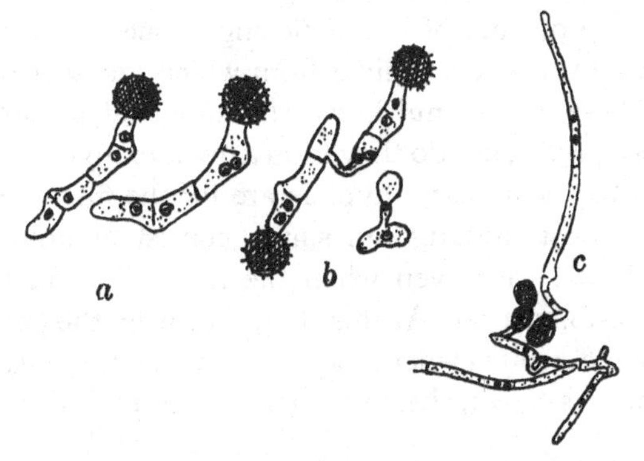


Fig. 153. *Ustilago Carbo*: *a*. formation of basidium; *b*. conjugation; *c*. binucleate mycelium; after Rawitscher.

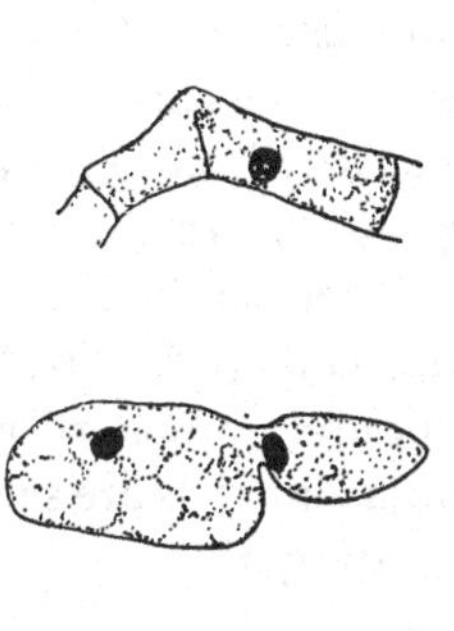

Fig. 154. *Ustilago Hordei*; conjugation; after Lutman.

nuclei, as well as the greater part of the cytoplasm, pass into it, leaving the conjugating cells comparatively empty. In these varieties of *U. Carbo* Lutman found that, after conjugation, the two nuclei lie closely pressed together so that it was sometimes impossible to differentiate them.

*Ustilago Tragopogonis pratensis* is parasitic on *Tragopogon pratensis*, in the flower heads of which it produces a mass of dark violet spores. In the young flower buds hyphae are abundant only in the anthers and ovary. Later they spread to the surface of these organs and form a dense mycelium of delicate filaments. According to Dangeard and to Rawitscher they divide, with the onset of spore-formation, into small binucleate cells; nuclear fusion takes place and the spore acquires a thick reticulate wall. In germination a three or four celled basidium is produced, each cell containing a single nucleus, and gives rise to uninucleate basidiospores, which increase by budding.

Federley, in 1903, described specimens of this fungus in which conjugation is followed not only by the migration of the nucleus of one of the cells concerned, but also by nuclear fusion (fig. 155). In view of the fusion in the young spore recorded by Dangeard and by Rawitscher the details of development in this species demand further investigation.

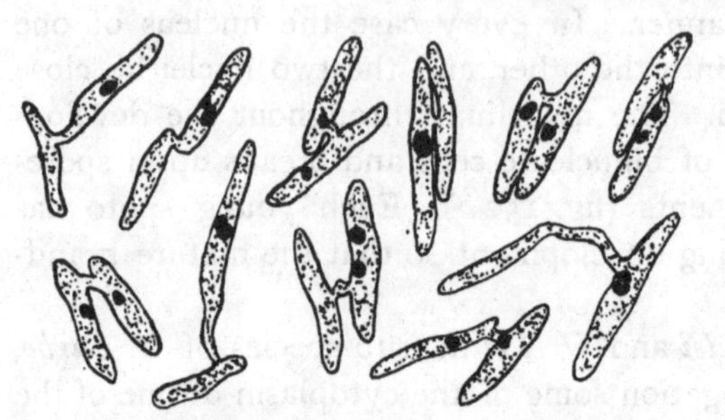

Fig. 155. *Ustilago Tragopogonis pratensis* (Pers.) Wint.; conjugation and nuclear fusion; after Federley.

*Ustilago Maydis*, the smut of *Zea Mays*, induces considerable hypertrophy. The deformations contain a mass of gelatinous mycelium from which brand-spores are produced. When mature, the spore mass causes the rupture of the enclosing tissues, and the spores escape. They germinate to produce basidia from which uninucleate basidiospores are abstricted. These in turn multiply by budding, but, according to Rawitscher, they never conjugate, nor do they form a definite mycelium (fig. 156 *a*). In the infection of the host plant, hyphae are for the first time developed, and, unlike those of most investigated smuts, consist of uninucleate cells (fig. 156 *b*). This is the case even when the hyphae begin to break up in preparation for spore-formation. At this stage, however, the ends of adjacent cells are seen to become swollen where they are in contact, the wall separating their protoplasm breaks down, the two nuclei come together in

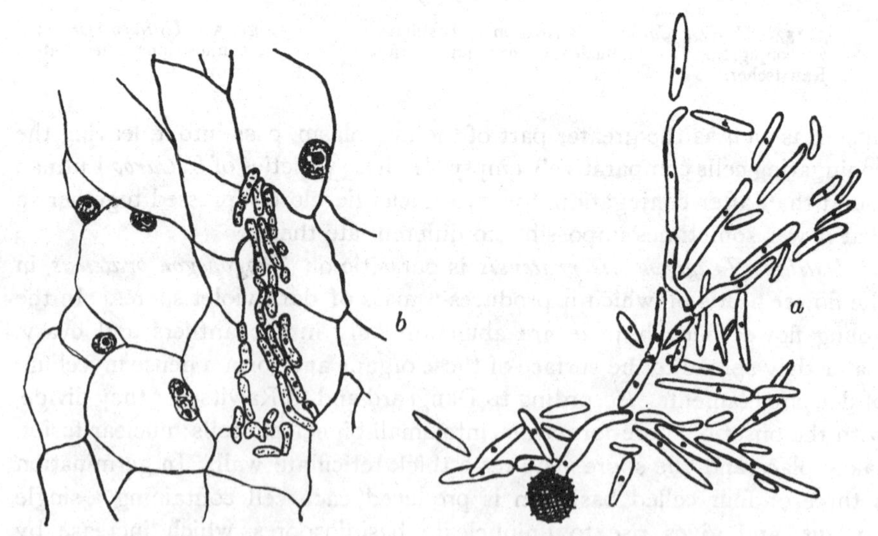


Fig. 156. *Ustilago Maydis*; *a*. basidiospores, ×540; *b*. uninucleate mycelium, ×420; after Rawitscher.

the swollen region and ultimately fuse (fig. 157). Thus the mature brand-spore in *Ustilago Maydis*, as in other species, contains a single fusion nucleus. Here, however, the nuclear association which usually takes place in the basidiospore is postponed till just before spore-formation. The parasitic mycelium is therefore haploid instead of diploid as in the majority of cases.

A very similar state of affairs has been described by I. Massee in *Ustilago Vaillantii* which attacks various liliaceous plants, hibernating in the bulb and forming spores in the anthers and ovary. In these organs the hyphae produce numerous short branches divided by transverse septa into cuboid cells which, like the cells of the vegetative mycelium, contain each a single nucleus. Alternate septa disappear by deliquescence so that binucleate segments are formed in each of which the two nuclei approach one another and fuse to form the single nucleus of the spore.

Germination in water takes place in the usual way; four-celled basidia are produced and give rise to basidiospores. Among these there is no evidence of conjugation.

*Ustilago antherarum* forms its spores in the stamens of members of the Caryophyllaceae; pollen-formation is inhibited, and the anthers become filled with fungal spores, which are distributed by the mechanism prepared for the dispersal of the pollen. Germination takes place in dung decoction with great readiness, the tubes being put out in three or four hours. Harper has observed that when the spore nucleus undergoes mitosis one of the daughter nuclei remains in the spore while the other passes into the basidium (fig. 158*a*). Here it divides, the two resultant daughter nuclei are separated by a wall, the nucleus remote from the spore divides again and a second wall is formed. Thus the three-celled basidium characteristic of this species is constituted.

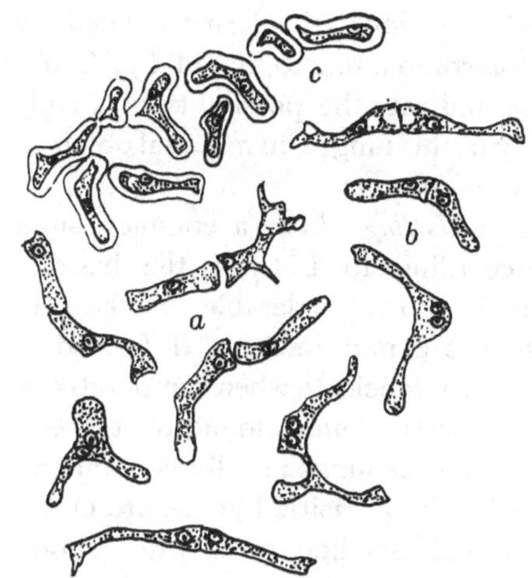


Fig. 157. *Ustilago Maydis*; *a*. uninucleate cells before spore-formation; *b*. conjugation; *c*. young, uninucleate brand-spores; after Rawitscher.

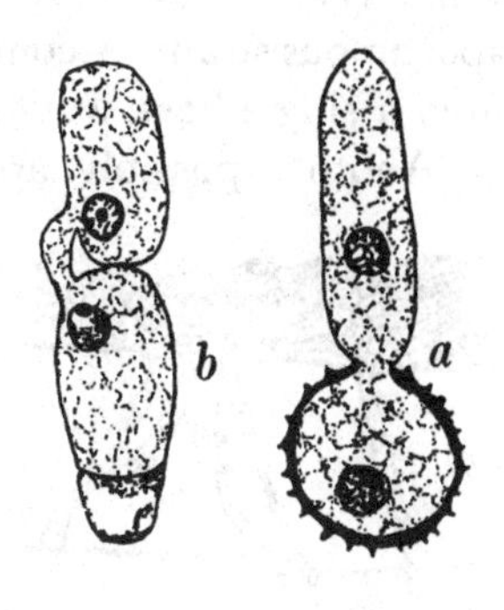


Fig. 158. *Ustilago antherarum* Fr; *a*. germination of brand-spore; *b*. conjugation; after Harper.

Basidiospores are budded off in abundance from all three cells, and in turn give rise to conidia. In the meantime the basidium has separated from its parent brand-spore, and the spore, after nuclear division, may produce another basidium, and others in succession at the same spot, so that free basidia accumulate in the culture. If cultures in nutrient solution are allowed to starve, association now takes place between basidial cells, basidiospores or conidia by means of conjugating tubes (fig. 158 *b*). The paired cells increase markedly in volume, but no interchange of cytoplasm takes place and the nuclei remain in their respective cells without visible change. Harper observed that when fresh beerwort was supplied to his cultures at this stage the production of conidia began again. They are produced from one or both of the conjugating cells, but only a single nucleus is concerned in the development of each conidium, the other remains quiescent and the conidia are uninucleate.

In a host plant, the name of which is not recorded, Werth and Ludwig failed to find binucleate elements, the youngest cells in which they could identify the nuclei being uninucleate (fig. 159 *a*). They infer that in this species nuclear association fails to take place, and no binucleate stage exists. This hypothesis accords well with Harper's observations on the saprophytic phase which he studied in material grown on *Lychnis alba*. On the other hand, a binucleate stage was identified by him in the sporogenous cells of *U. antherarum* on *Saponaria*, and by Dangeard on *Lychnis dioica*. These facts suggest the possibility of two or more varieties of *U. antherarum* on different hosts and differing in their cytological behaviour; the forms studied by Dangeard and Harper point to a condition comparable to that of *U. Maydis*, while Werth and Ludwig's observations indicate the possibility of a truly apogamous strain. A complete life-history of this fungus, in material obtained from a single host, would probably prove of interest.

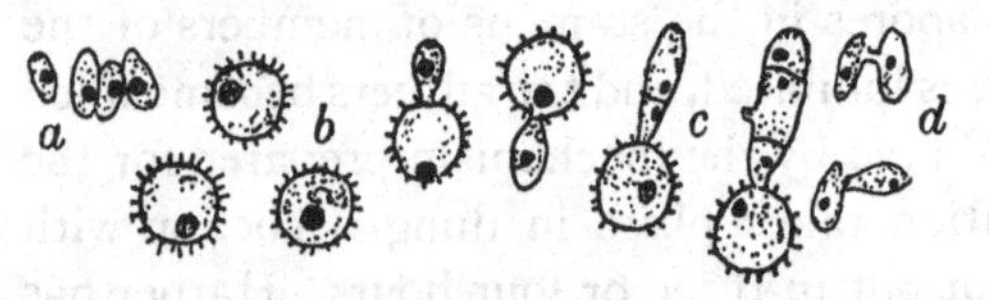

Fig. 159. *Ustilago antherarum* Fr.; *a*. young brand-spores · *b*. older brand-spores; *c*. basidia; *d*. basidiospores; after Werth and Ludwig

Another possibly apogamous form is *Ustilago levis*, a common smut on oats. According to Lutman the basidiospores give rise to considerable numbers of conidia. These are multinucleate if formed in crowded masses, uninucleate when comparatively isolated. Germinated conidia found on the epidermis of infected seedlings usually contain two or three nuclei. The parasitic hyphae are multinucleate and their swollen ends, when spore-formation is about to take place, contain ten to fifteen nuclei. The final segments however are

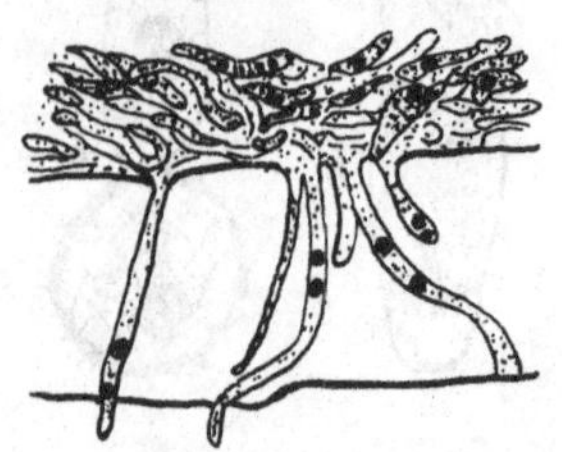
Fig. 160. *Ustilago levis* (K. and S.) Magn.; mycelium with multinucleate and binucleate cells; after Lutman

uninucleate or binucleate (fig. 160), but it is not known whether fusion takes place in them. The multinucleate character of the mycelial cells strongly suggests that no preliminary pairing of the nuclei occurs.

In *Ustilago Zeae* Lutman also observed a mycelium of multinucleate cells; at the time of spore-formation binucleate and uninucleate cells and finally uninucleate spores appear.

### *Tilletiaceae*

The principal genera of the Tilletiaceae are *Tilletia*, *Entyloma*, *Tuburcinia*, *Urocystis* and *Doassansia*. They have in common the continuous basidium with a terminal group of spores.

*Tilletia Tritici* and *T. foetens* are the stink brands of wheat, so called by reason of the strong odour of trimethylamin or herring brine given out by the brand-spores. The two species differ in the character of the epispore which is smooth in *T. foetens*, reticulate in *T. Tritici*. In both cases spores are produced in the ovaries of the host, all tissues of which, except the outer coat, are destroyed. The spore masses are garnered with the crop, and damage all grains with which they are threshed or ground. The infected flour and contaminated chaff and straw are causes of disease in man and animals.

On the germination of the brand-spore of *T. Tritici* the nucleus passes into the basidium and divides three times so that eight nuclei are formed. Eight basidiospores are budded off in a bunch at the apex of the basidium, and each receives a single nucleus. Frequently additional nuclear divisions take place and ten, twelve, or sixteen uninucleate spores may be produced. When the spores are fully formed short conjugating tubes grow out and connect neighbouring spores, often while these are still attached to the basidium (fig. 161).

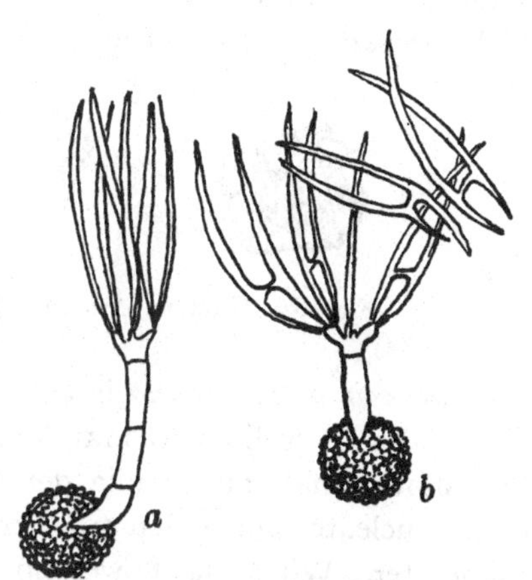


Fig. 161. *Tilletia Tritici* (Bjerk) Wint.; *a*. basidium thirty hours after germination of brand-spore; *b*. conjugation of basidiospores; × 300; after Plowright.

According to Rawitscher the nucleus of one cell of the pair passes over into the other and the nuclei lie near one another but without fusion. After conjugation the spores may become septate; from those which contain two nuclei filaments of binucleate cells grow out, and may give rise to conidia which are also binucleate. Under suitable conditions the binucleate hyphae bring about infection by pushing between the cells of the host seedling. In the cells of the sporogenous mycelium fusion of the pairs of associated nuclei takes place. Rawitscher observed a quite similar life-history in *T. laevis*.

In the parasitic mycelium of *Doassansia Alismatis* and *Entyloma Glaucii* (fig. 162) Dangeard observed binucleate cells and the fusion of their nuclei

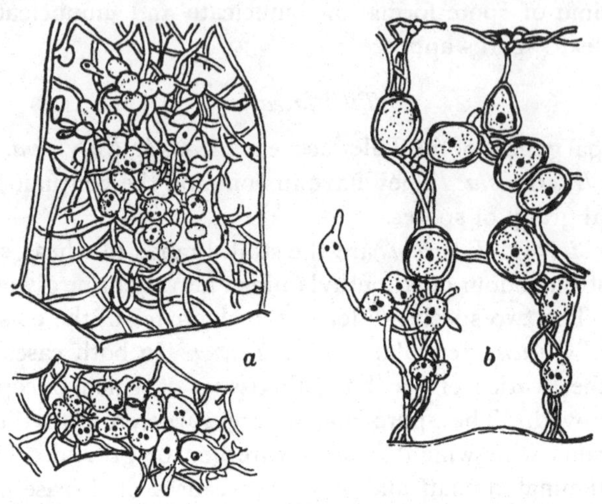


Fig. 162. Development of brand-spores; *a. Doassansia Alismatis* (Nees) Corn.; *b. Entyloma Glaucii* Dang.; after Dangeard.

in pairs in preparation for the formation of the brand-spores. The same stages were recorded by Lutman in *Doassansia deformans*, *Entyloma Nympheae* and *Urocystis Anemones* (fig. 163).

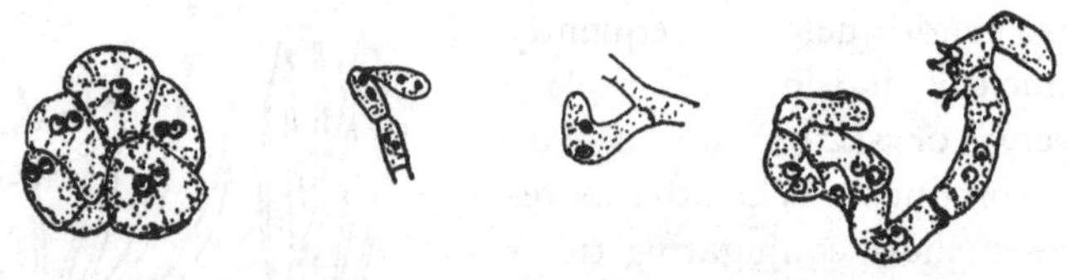

Fig. 163. *Urocystis Anemones* (Pers.) Wint.; mycelium and young spore ball; after Lutman.

*Tuburcinia primulicola* infects various species of *Primula* and gives rise to conidia as well as to brand-spores during its parasitic stage. Wilson has shown that the conidia develop in the young flower on a mycelium of uninucleate cells which apparently persists in the host plant throughout the winter. When the flower opens the conidia conjugate in pairs, and a nucleus passes through the connecting tube so that one conidium is empty and the other binucleate. Germ-tubes with paired nuclei are later produced and doubtless give rise to the mycelium of binucleate cells which bears the brand-spores. This mycelium ramifies in the superficial tissue of the placenta and between the ovules, giving rise to brand-spores in the same flowers in which the conidia were previously developed. On the germination of the brand-spores, basidia and basidiospores appear, but no conjugation could be observed. *T. primulicola* thus resembles the Uredinales in having two parasitic generations, the one with uninucleate, the other with binucleate cells. Its hibernating, uninucleate mycelium also recalls the parasitic mycelium of *Ustilago Maydis*, or of *U. Vaillantii*, which consists of uninucleate cells.

## USTILAGINALES: BIBLIOGRAPHY


1807 PRÉVOST, B. Mémoire sur la cause immédiate de la Carie. Fontanel, Montauban.

1847 BERKELEY, M. J. Propagation of Bunt. Trans. Hort. Soc. London, ii, p. 113.

1847 TULASNE, L. R. and C. Mémoire sur les Ustilaginées comparées aux Urédinées. Ann. Sci. Nat. 3 sér. vii, p. 12.

1853 DE BARY, A. Untersuchungen über die Brandpilze und die durch sie verursachten Krankheiten der Pflanzen. Müller, Berlin.

1854 TULASNE, L. R. and C. Second Mémoire sur les Urédinées et Ustilaginées. Ann. Sci. Nat. 4 sér. ii, p. 113.

1867 FISCHER von WALDHEIM, A. Sur la structure des spores des Ustilaginées. Bull. de la Soc. Imp. des Nat. de Moscou, xl, p. 242.

1883 BREFELD, O. Botanische Untersuchungen über Hefenpilze. V, die Brandpilze. Felix, Leipzig.

1887 ZU SOLMS LAUBACH, H. *Ustilago Treubii* Solms. Ann. du jard. bot. de Buitenzorg, vi, p. 79.

1888 WARD, H. MARSHALL. The Structure and Life History of *Entyloma Ranunculi* (Bonorden). Phil. Trans. clxxviii, p. 173.

1889 PLOWRIGHT, C. B. A Monograph of the British Uredineae and Ustilagineae. Kegan Paul, Trench & Co., London.

1894 DANGEARD, P. A. Recherches histologiques sur la famille des Ustilaginées. Le Botaniste, iii, p. 240.

1899 HARPER, R. A. Nuclear Phenomena in certain stages in the development of the Smuts.. Trans. Wisconsin Acad. Sci. Arts and Letters, xii, p. 475.

1904 FEDERLEY, H. Die Copulation der Conidien bei *Ustilago Tragopogi pratensis* Pers. Öfversigt af Finska Vetensk. Soc. Förhandlingar, lxvi, No. 2, p. 1.

1910 McALPINE, D. The Smuts of Australia. J. Kemp, Melbourne.

1911 LUTMAN, B. S. Some Contributions to the Life-history and Cytology of the Smuts. Trans. Wisconsin Acad. Sci. Arts and Letters, xvi, pt. III, p. 1911.

1912 RAWITSCHER, F. Beiträge zur Kenntniss der Ustilagineen. Zeitschr. f. Bot. iv, p. 673.

1912 WERTH, E. and LUDWIG, K. Zur Sporenbildung bei Rost und Brandpilzen. Ber. d. deutsch. Bot. Ges. xxx, p. 522.

1914 MASSEE, I. Observations on the Life-history of *Ustilago Vaillantii*, Tul. Journ. Econ. Biol. ix, p. 7.

1914 RAWITSCHER, F. Zur Sexualität der Brandpilze *Tilletia tritici*. Ber. d. deutsch. Bot. Ges. xxxii, p. 310.

1915 WILSON, M. The Life-History and Cytology of *Tuburcinia primulicola* Rostrup. B. A. Report, Sect. K. 1915.

# CHAPTER VIII

## PROTOBASIDIOMYCETES

### Uredinales

THE rust-fungi, members of the group Uredineae, Uredinales or Aecidiomycetes, including over 1700 species, are without exception obligate parasites on the stems, on the sporophylls and especially on the leaves of vascular plants, usually on those of angiosperms or gymnosperms but in one or two cases of ferns.

The mycelium ramifies in the tissues of the host, sends haustoria into the cells, and may act as a local stimulant causing more or less marked hypertrophy and consequent curling or malformation of the infected part. Starch may be stored by the host, and this is so abundant in the hypertrophies caused by the aecidial mycelium of *Puccinia Caricis* on the nettle, *Urtica parvifolia*, that they are eaten by the Himalayans; one or two other species are similarly employed. Where the mycelium penetrates into the perennial tissues of the host it is itself perennial.

**Spores and Sori.** On the mycelium several kinds of spore are produced, minute spermatia in spermogonia, aecidiospores in aecidia, uredospores and teleutospores, sometimes mixed, sometimes separate, in more or less definite sori. One or more of these types of spore may be lacking, but the teleutospores are almost invariably present, and it is on them that the classification of the group depends.

Naturally enough it was some time before the various types of spore were recognized as belonging to the same fungus and the old generic names

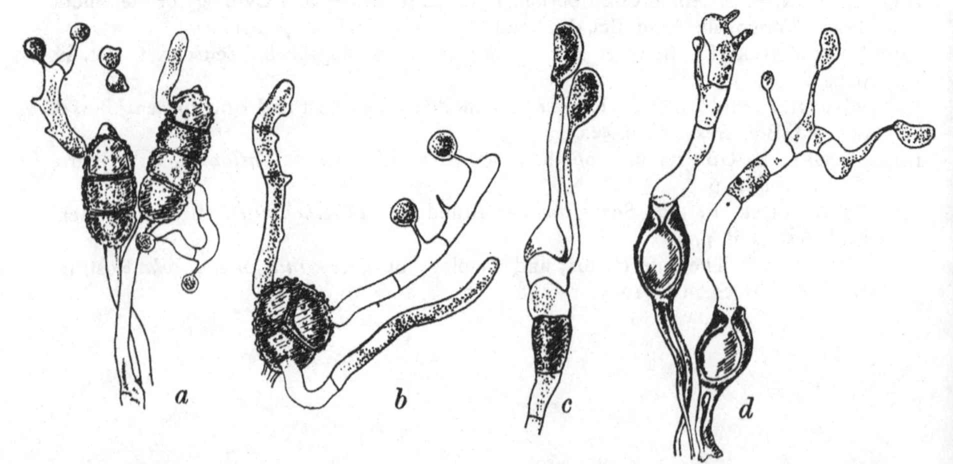


Fig. 164. Germinating teleutospores; *a. Phragmidium bulbosum* Schm.; *b. Triphragmidium Ulmariae* Lk.; *c. Coleosporium Sonchi* Lev.; *d. Uromyces appendiculatus* (*Fabae*) Lev.; after Tulasne.

of the spore forms other than the teleutospore, such as *Aecidium*, *Caeoma* and *Uredo*, still survive in our nomenclature.

The **teleutospores** (figs. 164, 165, 166) may be unicellular or they may be made up of two or more cells forming a compound structure, each cell of

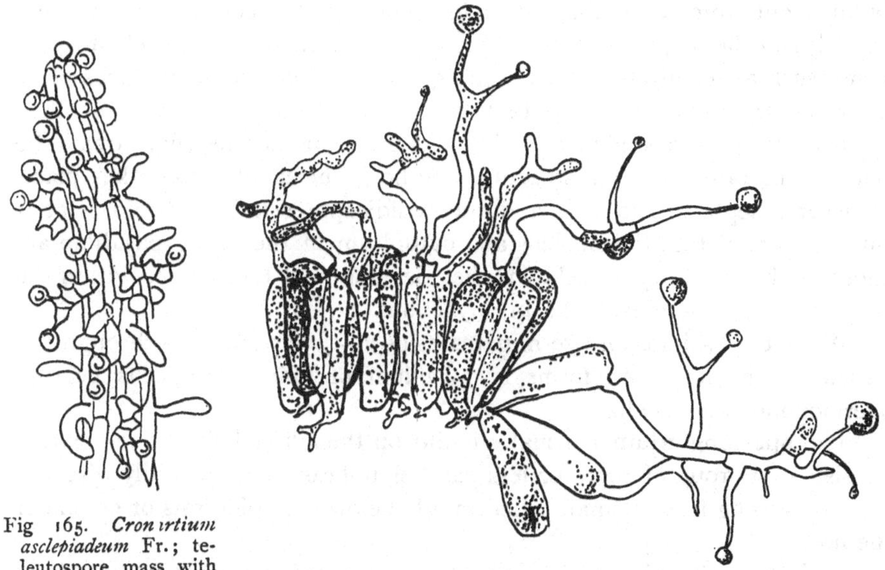

Fig 165. *Cronartium asclepiadeum* Fr.; teleutospore mass with basidia and spores; after Tulasne.

Fig. 166. *Melampsora betulina* Desmaz.; germinating teleutospores; after Tulasne.

which germinates independently. The teleutospore is simple in *Uromyces*, *Coleosporium*, and *Melampsora*, it is two-celled in *Gymnosporangium* and *Puccinia*, it is built up of three to ten superposed cells in *Phragmidium*, and of a larger number in *Xenodochus*. In *Triphragmidium* it consists of three cells laterally placed and in *Chrysomyxa* and *Cronartium* the simple teleutospores are so massed together as to simulate compound forms; their real nature is revealed by their early separation one from another. One-celled teleutospores occur exceptionally in the two-celled species and are known as **mesospores**.

The teleutospores may be massed together and incrusted in the tissues of the host, or they may be detached readily from their stalks and carried by the wind or by other agencies. Further development may take place as soon as conditions are favourable, or may be delayed till after a resting period, usually till the spring following development.

In either case the nucleus in each cell ultimately undergoes two successive divisions, which constitute a meiotic phase, and the daughter nuclei are separated by transverse walls, so that four uninucleate cells are produced. The teleutospore-cell thus functions as a tetrasporangium and divides into four portions, constituting the transversely septate **basidium.** From each cell a short, pointed branch or **sterigma** arises, its end dilates, a **basidiospore** (sporidium) is formed and receives the nucleus and cytoplasm of the cell from which it arose.

In *Coleosporium*, *Ochropsora*, and *Chrysospora*, nuclear division and septation take place within the teleutospore wall, and the basidiospores are budded out from it, so that the teleutospore cell becomes the basidium directly; in the majority of cases, however, the structure of the teleutospore is not such as readily to allow further growth, and development takes place after the extrusion of the contents as a tubular outgrowth, the so-called **promycelium**, surrounded only by a delicate membrane (fig. 167). The nucleus migrates into this structure and here nuclear division takes place, transverse septa are formed and the basidiospores are produced. But it must be noted that the nucleus and cytoplasm of the young basidium are those of the teleutospore cell, whether development takes place within the original wall or by means of a promycelium.

When the basidiospore germinates its germ-tube penetrates through the cuticle of the host and forms a mycelium of uninucleate cells bearing spermogonia and aecidia.

The **spermogonium** is usually found on the adaxial side of the leaf; it consists of a group of more or less parallel, unbranched, upwardly directed hyphae, arising from a small-celled tangle below the epidermis or cuticle of the host.

In the majority of cases the outer hyphae of each group elongate to form paraphyses, so that the spermogonium is restricted in extent, and acquires a flask-shaped or pyriform outline; the paraphyses push up through the ruptured epidermis of the host to project at a narrow ostiole (fig. 168 *b*). In

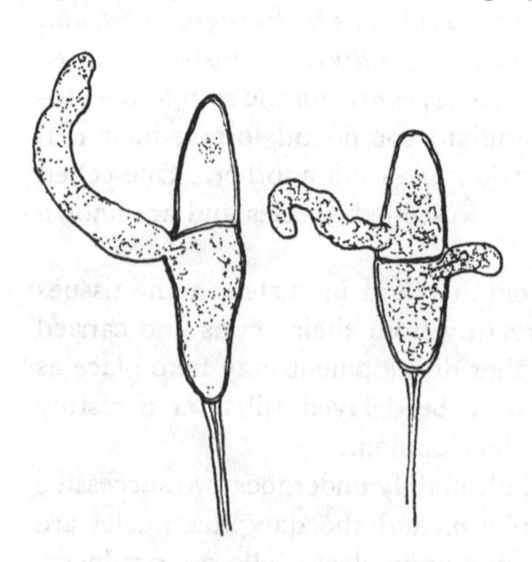

Fig. 167. *Gymnosporangium clavariaeforme* Rees; germinating teleutospores; ×666.

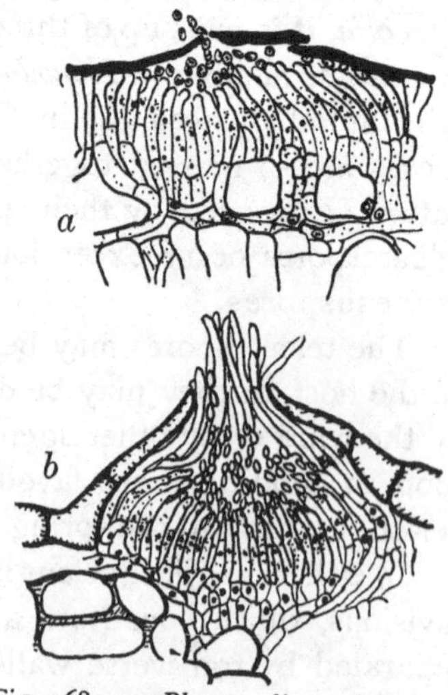


Fig. 168. *a. Phragmidium violaceum* Wint., ×330; *b. Gymnosporangium clavariaeforme* Rees, ×260; spermogonia; after Blackman.

simpler forms, such as *Phragmidium*, the spermogonium is indefinite in extent, and consists of spermatial hyphae arranged beneath the cuticle, which is perforated in the centre of the mass to form an ostiole. No regular paraphyses are produced but a few spermatial hyphae may elongate and project as sterile threads (fig. 168 *a*).

The **spermatial hypha** is a long, narrow cell with a central elongated nucleus. It is furnished at its free end with a ring of thickening which may be concerned with the disjunction of the spermatium. The development of the latter begins by the pushing out of a finger-like projection at the apex of the parent hypha. When it has attained its full size the nucleus of the hypha divides and one of the daughter nuclei enters the spermatium, which is cut off by a wall formed just above the thickened ring (fig. 169).

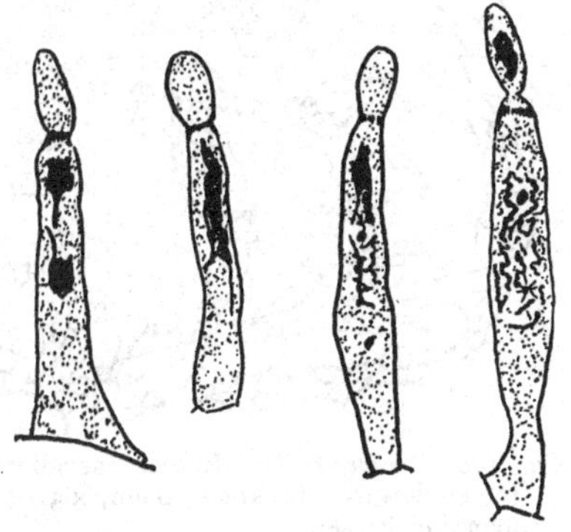

Fig. 169. *Gymnosporangium clavariaeforme* Rees; development of spermatia, × 1185; after Blackman.

The mature **spermatium** is a small more or less oval cell, enclosed in a very thin wall. The cytoplasm is finely granular with apparently no reserve material, and the nucleus is of relatively large size. When cultivated in solution of sugar or honey, spermatia have been induced to undergo a form of yeast-like budding, and this has been observed under natural conditions by Robinson in *Puccinia Poarum*. But, though many attempts have been made, it has so far proved impossible to bring about the formation of a mycelium. It seems, therefore, pretty clear that the spermatia are useless as agents of infection, and they differ also in structure from ordinary asexual spores. On the other hand the suggestion was long ago made that they may be male reproductive elements, and this is borne out by their large nuclei and lack of reserve material, and is by no means invalidated by the fact that they possess some slight power of germination. Recent investigation has shown that they are now no longer functional.

As a rule considerable numbers of spermatia are to be found in various stages of degeneration scattered around the ostioles of the spermogonia. In some cases the spermatia are aggregated in sticky masses and appear to attract insects. The presence of sugars in the spermogonial contents has been demonstrated for species of *Uromyces*, *Puccinia*, *Endophyllum*, and *Gymnosporangium*; in some cases the spermogonia also possess a strong odour as in *Puccinia suaveolens*, or occur on bright spots which contrast with the green of the surrounding tissue. Such spots are usually yellow or orange, but are white in *Uromyces Fabae* and reddish-purple in *Puccinia*

*Phragmitis.* A corresponding discoloration takes place around the young aecidia, and there is thus some suggestion that the spermatia, when functional, were carried to their destination by insects.

The **aecidia** occur in groups, usually on the abaxial side of the leaf; in them the **aecidiospores** are produced in basipetal rows (fig. 170) alternating with small, abortive, **intercalary cells**, by the disintegration of which they are set free. They may be carried to considerable distances by the wind, and there is evidence that they are sometimes distributed by means of insects or of snails. The mature aecidiospore is usually subglobose or polygonal in form, it is enclosed in a thick wall perforated by several germ-pores, and contains red, yellow or orange pigment, and always two nuclei. In germination a hypha is put out which enters the host plant through one of the stomata and so penetrates into the intercellular spaces.

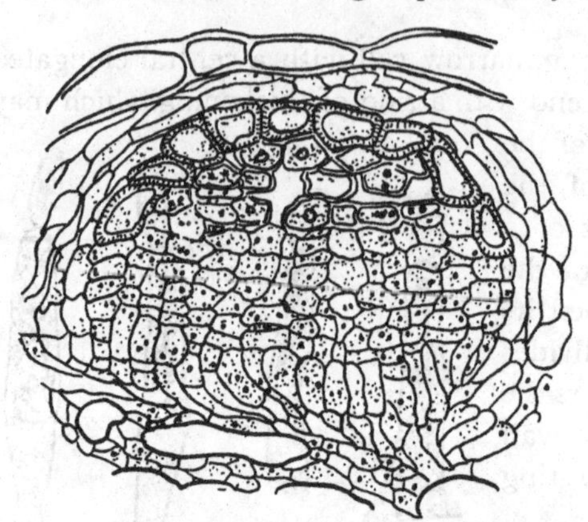

Fig. 170. *Uromyces Poae* Raben.; aecidium just before the epidermis is broken through, × 310; after Blackman and Fraser.

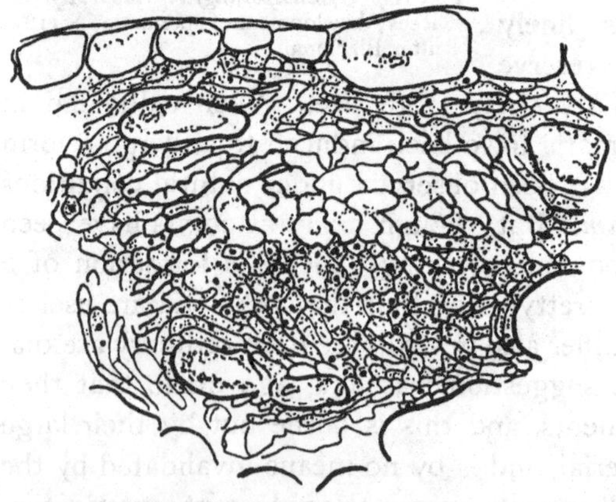

Fig. 171. *Uromyces Poae* Raben.; young aecidium, × 370; after Blackman and Fraser.

The development of the aecidium begins by the massing of hyphae either deep in the tissues of the host (*Gymnosporangium clavariaeforme, Puccinia Poarum* (Blackman and Fraser '06), *Puccinia Falcariae* (Dittschlag '10)), or directly below the epidermis (*Phragmidium violaceum* (Blackman '04), *Uromyces Poae* (Blackman and Fraser '06) (fig. 171), *Puccinia Claytoniata* (Fromme '14)); these hyphae give rise to a more or less regular series of uninucleate cells. These are the **fertile cells**, but, before developing further, each, at any rate in the relatively primitive forms (caeomata), may cut off one or occasionally more terminal **sterile cells** which ultimately degenerate. The fertile cells may unite laterally in pairs (fig. 172), so that binucleate compound cells are formed; they may similarly pair with the cells below them (Fromme '14), or each may receive a second nucleus by migration from a neighbouring vegetative cell (fig. 173). In each case they now constitute the basal cells of the rows of spores and they proceed at once to cut off aecidiospore mother-cells, each of which in turn divides to separate a small intercalary cell below from the aecidiospore above.

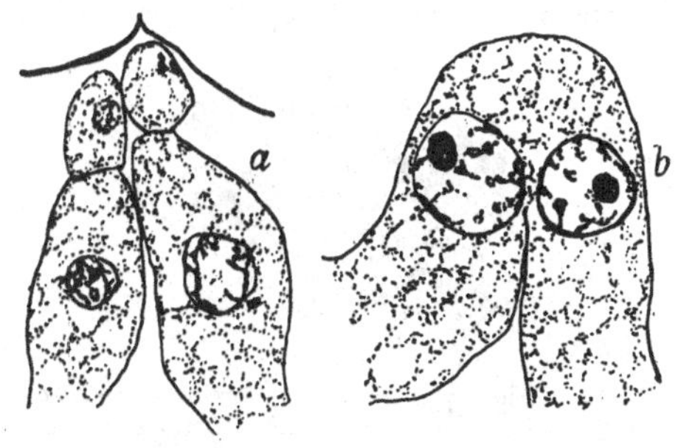


Fig. 172. *Phragmidium speciosum* Fr.; *a.* fertile and sterile cells; *b.* fusion of two fertile cells; after Christman.

Exceptionally binucleate cells may be observed before the fertile layer is differentiated. In *Puccinia Poarum* nuclear migrations sometimes take place between the vegetative cells at the base of the very young aecidium. These cells may grow up, either at once or after division, to form fertile cells.

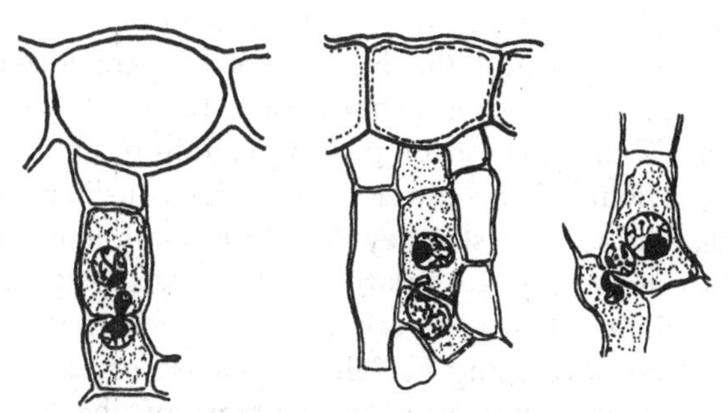

Fig. 173. *Phragmidium violaceum* Wint.; migration of second nucleus into fertile cell of caeoma, ×950; after Blackman.

The aecidiospores, then, are the products of a sexual process by means of which two nuclei become associated within the limits of a single protoplasmic mass, forming the **dikaryon** or **synkaryon** of Maire. The nuclei thus brought together do not fuse, but undergo simultaneous division (fig. 174), so that a daughter nucleus from each passes into every new cell. Conjugate division is continued when the aecidiospore germinates and a mycelium of binucleate cells is produced. The sporophyte of the rusts is thus normally inaugurated in the fertile cells of the aecidium.

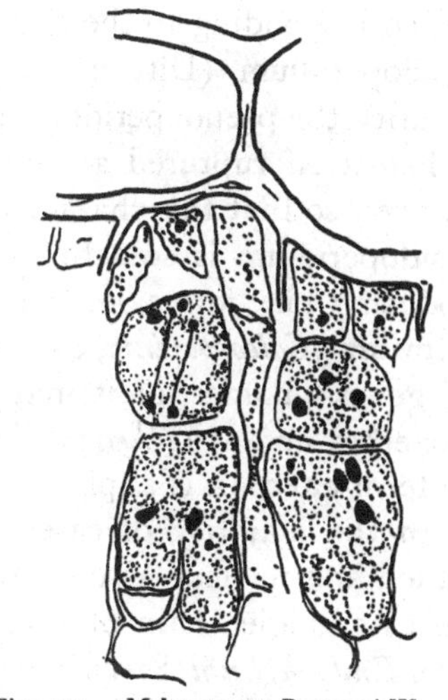

Fig. 174. *Melampsora Rostrupi* Wagn.; paired fertile cells, ×1200; after Blackman and Fraser.

It is not unusual to find spores and vegetative cells which contain three or more nuclei; in these, as in the binucleate cells, and, indeed, in multinucleate cells of many different groups of plants, conjugate division takes place (fig. 175). The origin of the trinucleate cell by the fusion of three fertile cells has been observed, and no doubt it may arise by the migration of a second vegetative nucleus.

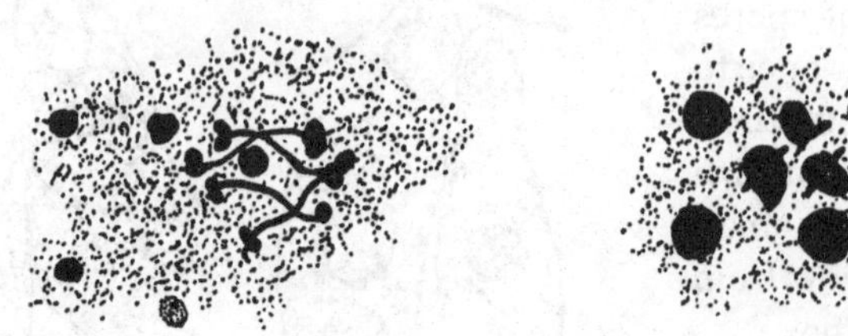

Fig. 175. *Puccinia Poarum* Niels.; conjugate division, × 2280; after Blackman and Fraser.

At the periphery of the aecidium the cells cut off from the basal cells divide in the usual way, so that two cells corresponding respectively to the aecidiospore and the intercalary cell are formed from each. The upper (aecidiospore) cells acquire very thick striated walls, lose their contents and form a sheath or **pseudoperidium** about the sporogenous part. The behaviour of the lower cells varies considerably; Kurassanow has shown that in some cases they are quite small, like typical intercalary cells, while in others they are relatively well developed and form part of the pseudoperidium. This is especially the case where the tissue to be broken through by the developing aecidium is dense or extensive. Centrally the pseudoperidium arches over the contents of the aecidium. In this region it is derived from the cells first cut off by the central basal cells. These, like the others, divide transversely, and one of the daughter cells, usually the outer one, corresponding to the aecidiospore, becomes one of the elements of the pseudoperidium (Dittschlag, Kurassinow). When the aecidium reaches maturity the pseudoperidium pushes through the epidermis of the host and is then itself ruptured and exposes the ripe spores. It becomes torn and recurved so that the characteristic cluster-cup is produced (fig. 176). The pseudoperidium is sometimes much elongated and cylindrical or inflated, producing the forms known as **roestelia** (*Gymnosporangium*), and **peridermium** (*Coleosporium*, *Cronartium* and allied genera), so-called from their old generic names, or it may be represented only by a few paraphyses or altogether absent (*Phragmidium*, *Melampsora*). The latter forms, to which the term **caeoma** is applied, are probably primitive.

In the majority of cases, after the fertile and sterile cells have been formed and nuclear association has taken place, the basal cells each give rise to a single chain of spores, but occasionally (*Puccinia Falcariae* (fig. 177), *Endophyllum Sempervivi*) they may branch and thus produce two or more spore-rows. In certain other species the basal cells regularly form a number of lateral buds or branches and each of these is cut off as a spore

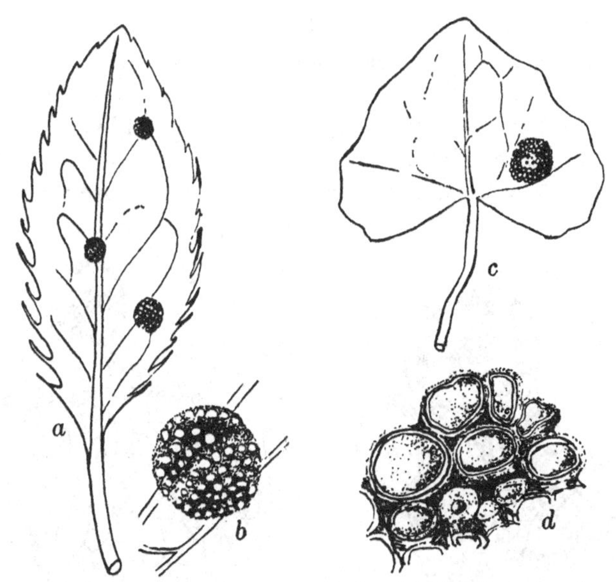


Fig. 176. *Puccinia Graminis* Pers.; *a.* infected leaf of *Berberis vulgaris*, nat. size; *b.* group of aecidia, × 5. *Uromyces Poae* Rabenh.; *c.* infected leaf of *Ranunculus ficaria*, nat. size; *d.* group of aecidia, × 20; E. J. Welsford del.

mother-cell (fig. 178). The spore mother-cell divides in the usual way, separating the aecidiospore above from its sister-cell below, but the latter here forms an elongated stalk instead of an intercalary cell. Each outgrowth of the basal cell thus produces only a single spore, the mode of formation of which is exactly similar to that of a uredospore. The fructification has generally been regarded as a uredosorus and is known as the **primary uredosorus,** in reference to its appearance relatively early in the season.

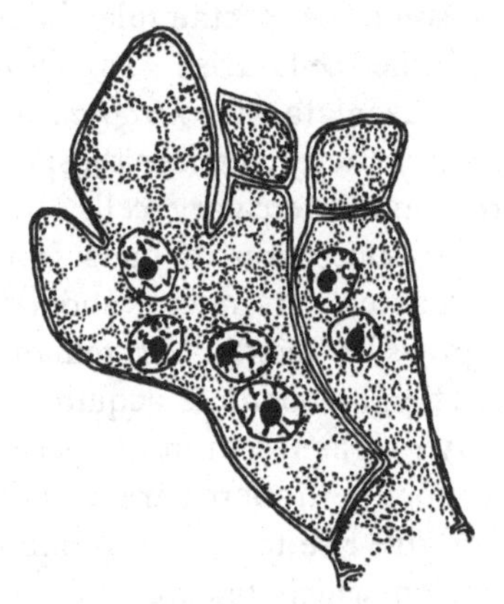

Fig. 177. *Puccinia Falcariae*; branched fertile cell of aecidium or primary uredosorus, × 1200; after Dittschlag.

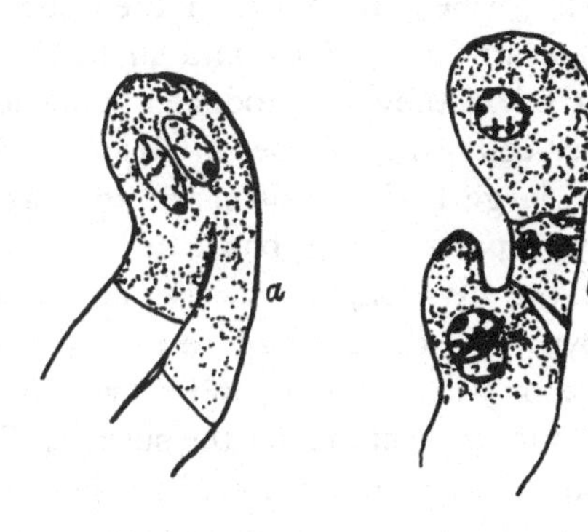


Fig. 178. *Phragmidium Potentillae-Canadensis* Diet.; *a.* conjugation; *b.* branched fertile cell; after Christman.

But the fact that these sori are developed on the same mycelium as the spermogonia, the fact that in their fertile cells nuclear association takes place and the fact that in the formation of the fertile cell a sterile cell is cut off, all suggest that the true homology is with the aecidium.

The mycelium formed by the germination of the aecidiospore grows with renewed energy. It consists of binucleate cells giving rise to **uredospores**. These are borne in groups or **uredosori** (fig. 179) which may be surrounded

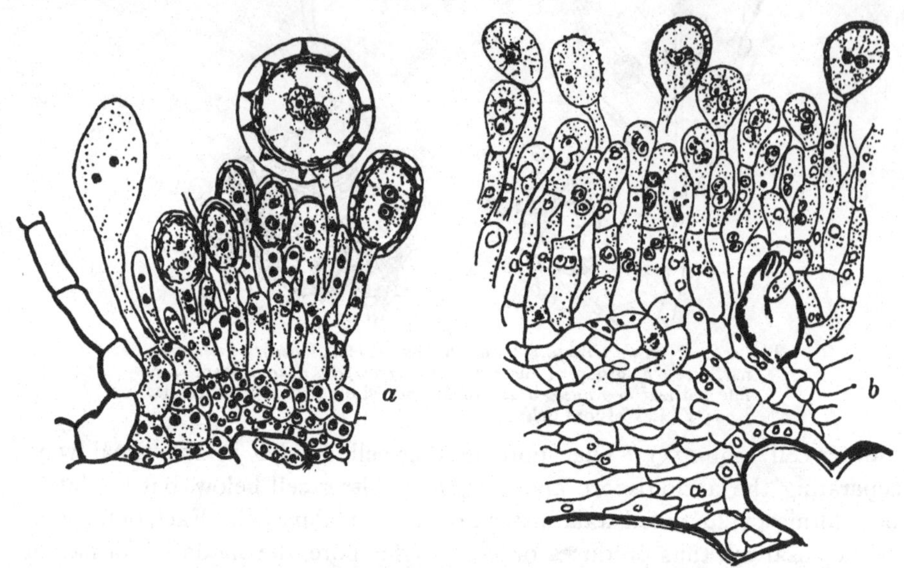


Fig. 179. *a. Phragmidium Rubi* Pers.; uredosorus, ×600; after Sappin-Trouffy; *b. Phragmidium violaceum* Wint.; uredosorus, ×480; after Blackman.

by paraphyses, or in certain genera (*Pucciniastrum*, *Uredinopsis*) by a pseudoperidium. In the young sorus a regular layer of somewhat rectangular basal cells is formed, from which the uredospore mother-cells arise. In *Coleosporium*, in *Chrysomyxa*, and in the secondary caeomata of *Phragmidium subcorticium*, they are produced in vertical rows like the typical aecidiospore mother-cells and divide to form uredospores and intercalary cells, but, in the large majority of cases, they appear as a succession of buds from different parts of the basal cell. Each bud elongates, its nuclei undergo conjugate division, a stalk is cut off which grows in length but remains narrow, while the uredospore enlarges considerably, its contents acquire an orange or yellow colour, and its wall is variously roughened in most species by minute projections on the surface. Two or more germ-pores are usually present and the uredospore, like the cells which give rise to it, is invariably binucleate; it produces a binucleate mycelium on which teleutospores or further crops of uredospores are formed.

Certain species (*Puccinia vexans*, etc.), occurring under very dry conditions, produce a second type of uredospore with thick walls which are adapted to survive unfavourable conditions; these are known as **amphispores.**

Both aecidio- and uredospores germinate readily and without a rest if fully ripe, but many are shaken off by wind and rain before they reach maturity and remain incapable of germination. Moreover it is stated that spores will not ripen properly on leaves that have been removed from the plant.

Sooner or later the mycelium of binucleate cells gives rise to **teleutospores**; these are characteristically grouped together in **teleutosori** (fig. 180),

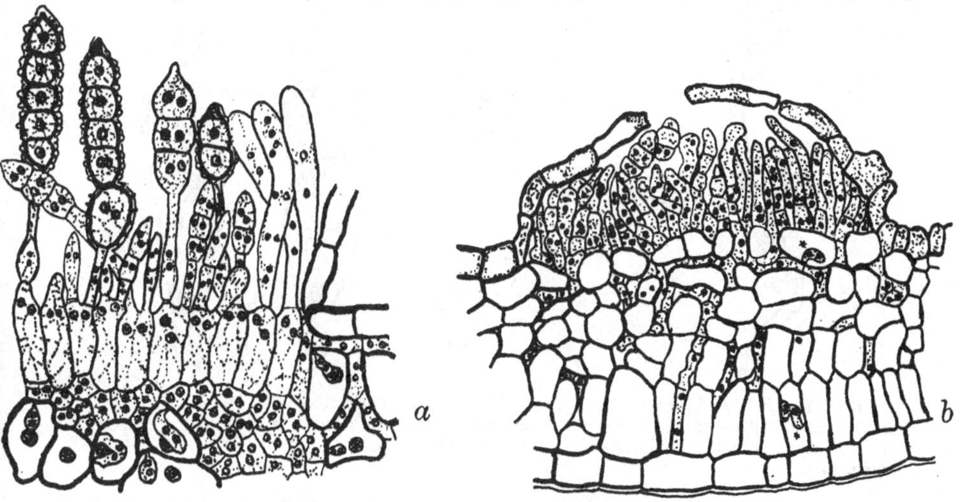


Fig. 180. *a. Phragmidium Rubi* Pers.; teleutosorus, × 240; after Sappin-Trouffy; *b. Phragmidium violaceum* Wint.; teleutosorus, × 240; after Blackman.

except in the genus *Uredinopsis*, on ferns, where they are scattered. Like the uredospores the teleutospores are with or without paraphyses and like them arise from rectangular basal cells. They appear as narrow binucleate outgrowths in which one or more divisions take place so that, in the majority of cases, a stalk is formed below and the simple or compound teleutospore is produced above (fig. 181). The stalk may increase considerably in length (*Gymnosporangium*, *Uromyces*, *Puccinia*) or may be very short or absent (*Coleosporium*, *Melampsora*).

As already stated the young teleutospore cell is binucleate (fig. 182); when the wall is fully thickened the two nuclei fuse and the spore passes into the resting state. On the renewal of its development two nuclear divisions occur and the gametophytic phase is initiated with the production of the uninucleate basidiospore.

In the Uredinales the fertilization process thus takes place in two stages, nuclear association being separated by a longer or shorter series of vegetative cells from nuclear fusion. We have here a difference in degree though not in kind from the normal process. In *Pinus sylvestris*[1] the male and female nuclei lie side by side but do not fuse till their chromosomes become mingled on the first spindle of the embryo; in many of the protozoa and in some other animals a series of conjugate divisions may precede the combination of the paternal and maternal chromosomes in a single membrane.

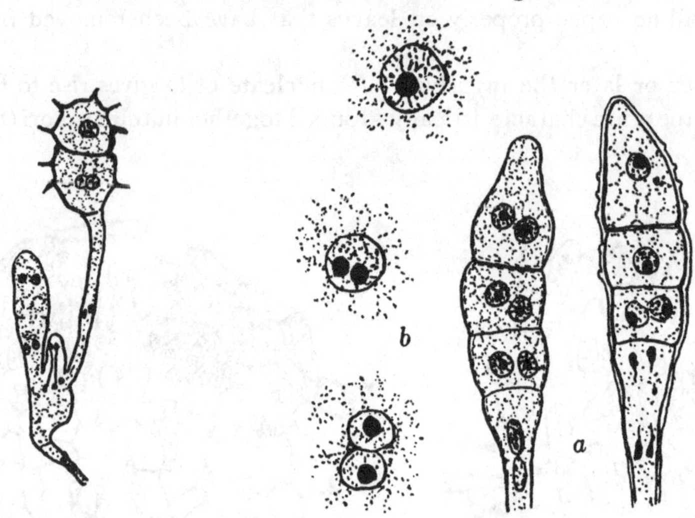


Fig. 181. *Puccinia Podophylli* S.; fertile cell of teleutosorus giving rise to teleutospores; after Christman.

Fig. 182. *Phragmidium violaceum* Went.; *a.* teleutospores, × 1080; *b.* fusion of nuclei in teleutospore, × 1520; after Blackman.

It may be hazarded that in the Uredinales the similarity of the physiological history of the nuclei before they become associated is responsible for a minimum of attraction between them, so that there is no sufficiently strong impulse towards fusion till meiosis is about to take place; being, however, in the same cell, they have no opportunity of dissolving partnership and the influences which bring about meiosis affect both alike.

A considerable similarity exists in the arrangement of the different groups of sporogenous cells. The uredo- and teleutosori are clearly comparable, both are of indefinite extent, with or without a border of paraphyses, and both consist of groups of rectangular basal cells from which the spore mother-cells arise in horizontal series and divide to produce the simple or compound spore and the stalk-cell. Sometimes, however, the uredospores are borne in vertical series, one below the other, and the sister-cells of the spore form short, intercalary cells instead of stalk cells (fig. 183).

This arrangement and that of the so-called primary uredospores link the uredosorus to the aecidium, suggesting the homology of the stalk and intercalary cells. In the simplest aecidia, those of the caeoma type, we have a

[1] Blackman, V. H., 1898, *Phil. Trans.* B. cxc, p. 395

group of basal cells of indefinite extent, from which the aecidiospore mother-cells are cut off. The aecidium is, in fact, no more a definite organ than the uredo- or teleutosorus, and only appears so in the more elaborate forms because of the modification of its peripheral cells to form a pseudoperidium. The important distinction lies not in the general morphology of the sorus, but in the fact that an association of two nuclei from different cells takes place in the basal cell of the aecidium and in the specialization indicated by the separation of the sterile cells.

Fig. 183. *Coleosporium Sonchi*; uredosorus, × 545; after Holden and Harper.

In its general structure the spermogonium, consisting as it does of a series of spermatial hyphae with or without circumjacent paraphyses, is not very different from the other sori, and, in the simplest cases, it also is of indefinite extent.

**Omission of Spore Forms.** In many rusts one or more spore forms are omitted; the case where the so-called primary uredospore is substituted for the typical aecidiospore has been already described; these and others in which the characteristic aecidium or caeoma alone is lacking are distinguished by the prefix **brachy**.

**Hemi-** indicates the presence of uredo- and teleutospores without aecidia or spermogonia.

The suffix **opsis** is used for forms with aecidia and teleutosori. They lack uredosori but a few uredospores are sometimes found in the teleutosorus.

**Micro-** and **lepto-** forms have teleutospores with a few occasional uredospores and sometimes spermogonia. The teleutospores germinate in the former group only after a period of rest, in the latter upon the same host plant as soon as they reach maturity.

Species with the full complement of spores are distinguished by the prefix **eu**.

As already pointed out, the sporophyte in some of the *brachy-* species starts from the fertile cells at the base of the so-called uredosorus, and this may very probably prove to be true in all cases where the ordinary aecidium is absent and spores developed like uredospores accompany the spermogonia.

The alternation of generations in the *-opsis* forms is also normal, for these are characterized by the omission of typical uredosori the development of which is related to no significant change in the nuclear life-history.

In *micro-* and *lepto-* forms the basidiospore germinates to produce, as in *eu-* species, a mycelium of uninucleate cells on which spermogonia may occasionally be borne. The mycelium becomes binucleate either during vegetative development (*Uromyces Scillarum*, *Puccinia Adoxae*) or below the young teleutosorus, and the fusion in the teleutospore takes place in the usual way. In the *micro*- form *Puccinia transformans* Olive observed that the binucleate condition was brought about by the fusion in pairs of cells to form the basal cells from which the teleutospores arose and the same has been reported by Moreau for *Puccinia Buxi* and *Uromyces Ficariae*. In *Puccinia Malvacearum* Moreau occasionally found a difference in size between the fusing cells, and Werth and Ludwig observed the migration of the nucleus of the smaller cell into the larger (fig. 184 *a*).

Below the teleutosorus of *Puccinia Podophylli* also, Christman found nuclear migrations in progress (fig. 184 *c*). Such cases clearly suggest that here, as in the mycelium below the aecidium of *P. Poarum* and in the prothallus of pseudapogamous ferns, the sporophyte is initiated by the association of two vegetative nuclei. Christman, however, observed that in certain cases migration took place between cells already binucleate, and he hence regards migrations in this case, and is inclined to regard all other migrations in rusts, as due to pathological causes.

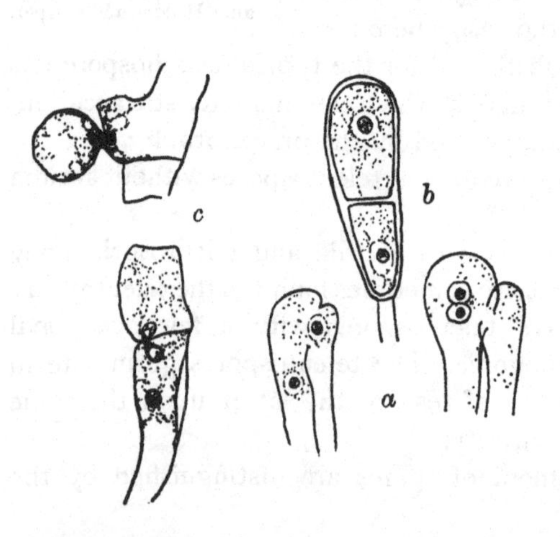


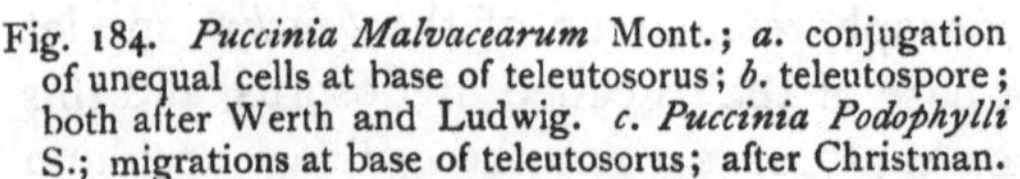
Fig. 184. *Puccinia Malvacearum* Mont.; *a.* conjugation of unequal cells at base of teleutosorus; *b.* teleutospore; both after Werth and Ludwig. *c. Puccinia Podophylli* S.; migrations at base of teleutosorus; after Christman.

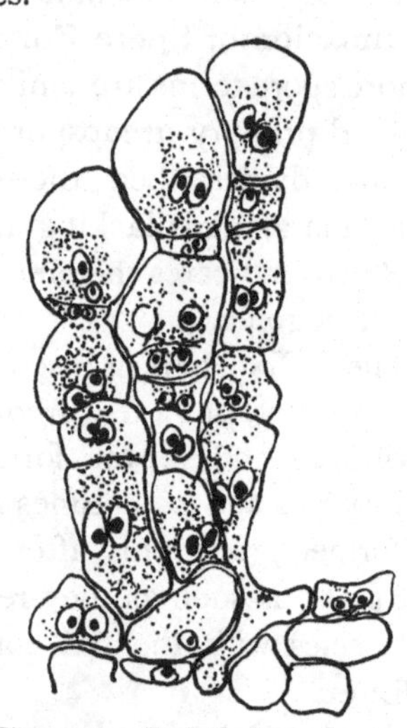
Fig. 185. *Endophyllum Sempervivi* Lev.; fertile cells and spores; after Hoffmann.

A sporophytic stage of exceptionally brief duration is also found in the species of *Endophyllum* and in the form on *Rubus frondosus* known as *Kunkelia nitens*[1]. In both cases the characteristic spores are developed in basipetal chains (fig. 185), and in both the fertile cells which give rise to them fuse in pairs (Olive '08; Hoffmann '11), so that the spore mother-cells, intercalary cells and spores receive two nuclei each. In *Endophyllum* the spores are enclosed in a pseudoperidium of barren cells so that the sorus appears as a typical aecidium, in *Kunkelia nitens* it is of simpler, indefinite

[1] *Kunkelia nitens* (Schwein.) Arthur = *Caeoma nitens* Burr.

form; in both cases it is borne in association with spermogonia on a mycelium of uninucleate cells. But the spore germinates like a teleutospore

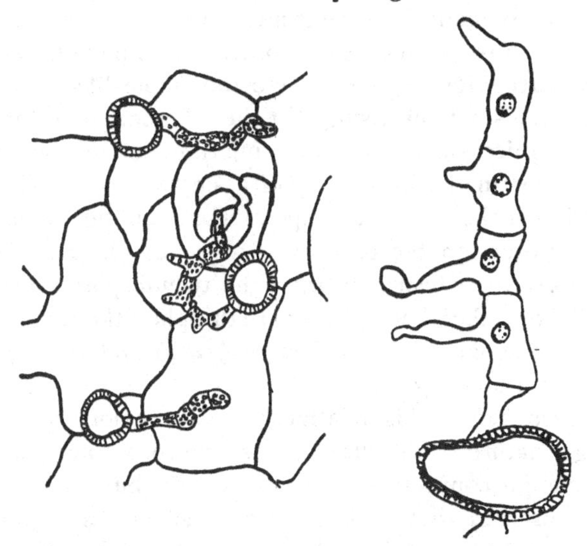

Fig. 186. *Endophyllum Sempervivi* Lev.; spores giving rise to basidia; both after Hoffmann.

(fig. 186); its two nuclei fuse (fig. 187), its contents are extruded as a promycelium, two successive nuclear divisions occur, cross walls appear and four basidiospores are produced, which, in due course, give rise to a uninucleate mycelium. The sporophytic stage thus endures only from the fusion of the fertile cells until the germination of the spores which they produce.

Incidentally these observations in the case of *Kunkelia nitens* have demonstrated that the caeoma of this fungus is not a stage in the life-history of the teleutospore-producing *Puccinia Peckiana* on the same host, for the mycelial cells of *P. Peckiana* are binucleate and the teleutospores germinate in the usual way.

The development of an apogamous aecidium has been observed by Moreau in a variety of *Endophyllum Euphorbiae* on *Euphorbia sylvatica*; here the basal cells, aecidiospores and cells of the pseudoperidium are uninucleate throughout their development, the aecidiospore germinates to form a promycelium of three or four cells and neither nuclear association

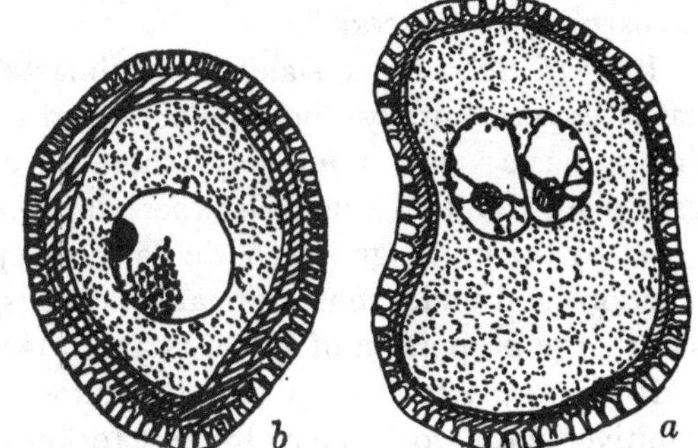


Fig. 187. *Endophyllum Sempervivi* Lev.; *a.* nuclear fusion in spore; *b.* synapsis in fusion nucleus; after Hoffmann.

nor nuclear fusion takes place at any stage, the diplophase being wholly omitted.

**Heteroecism.** In many rusts the gametophytic and sporophytic mycelia occur on different host plants. Such forms are termed **heteroecious** in contrast to the **autoecious** species where the whole life-history is passed on a single host. It is not surprising that the different spore forms on such species were recognized and described some time before it was understood that they are stages in the life-history of a single fungus. The final proof of the relationship of the aecidia and spermogonia on the one hand and the uredo- and teleutosori on the other, was given by de Bary in 1865 for *Puccinia Graminis*, the wheat rust or, as the teleutospore stage was called by early investigators, the wheat mildew. In this plant the haplophase occurs on the leaves of the Barberry (*Berberis vulgaris*) and the diplophase on wheat, oats, rye and other grasses.

Long, however, before this relationship was demonstrated, and even before the fungal nature of the disease was known, farmers had begun to suspect some malign connection between barberry bushes and their wheat crop, and had observed that dark areas of blackened and injured wheat were apt to occur in the neighbourhood of such plants. In the State of Massachusetts an act was passed requiring the inhabitants to extirpate all barberry bushes before a given date in 1760 and Marshall, of Norfolk, writing in 1781, says that "it has long been considered as one of the first of vulgar errors among husbandmen that the barberry plant has a pernicious quality (or rather a mysterious power) of blighting wheat which grows near it[1]." It is hardly to be wondered at that learned persons of the time repudiated this belief, or, as Marshall says of himself, "very fashionably laughed at it." It was not till 1797 that Persoon identified the wheat disease as a fungus and gave it the name it still bears. In 1805 Sir Joseph Banks called attention to its resemblance to the rust on the barberry, suggesting that it might be one and the same species, "and the seed transferred from the barberry to the corn."

In 1816 Schoeler, a Danish schoolmaster, set himself to deal with the matter experimentally, and applied rusted barberry leaves to some marked plants of rye; after a few days these were badly affected while not one rusty plant could be found elsewhere in the field. His discovery was confirmed by the investigations of de Bary, who performed the infection in both directions and under more critical conditions, and it has since been shown that a large proportion of the rusts are in fact heteroecious.

It seems pretty evident that the autoecious condition must have been primitive and it would be of interest to know what factors determined the adoption of different hosts for the different phases of the life-history.

[1] *Rural Economy of Norfolk*, 2nd ed. vol. ii, p. 19, London, 1795.

Christman and Olive have inferred that the ancestral type of the heteroecious species was a form with teleutospores only, a *lepto-* or a *micro-* form occurring on the host of the present gametophyte. Other investigators (Tranzschel '04, Grove '13) who regard the aecidium as an essential constituent of the primitive rust, have suggested that heteroecism may have arisen in relation to host plants with a short vegetative period. In such a case there would hardly be time for the production of the full complement of spores and the fungus might either shorten its elaborate life-history, giving the *micro-* or similar forms (*Uromyces Scillarum* on wild hyacinth, *Puccinia fusca* on anemone), or some of the spores might become adapted to life on a new host. This might be the case in particular with the aecidiospores, the development of which, owing perhaps to the recent fertilization stage, is especially vigorous. Aecidiospores must fall on hundreds of leaves besides those of the host, and the germ-tubes in their case enter through the stomata. If then an aecidiospore germinated and penetrated a satisfactory host it is suggested that a mycelium might develop and further adaptations might fix the heteroecious habit. Again it is readily understandable that the Gramineae and other hosts with similarly refractory cuticles are easily infected by the germ tubes from the aecidio- and uredospores which pass through the stomata but not by those of the basidiospores which, in a large majority of cases, penetrate the walls of the epidermal cells. This fact may be significant in relation to the return of the parasite to its gametophytic host each spring.

There is reason to believe that some species have an autoecious and a heteroecious variety and the study of such forms is likely to prove of great assistance.

**Specialization of Parasitism.** The parasitism of the rusts shows very marked specialization so that biological species have arisen, which, though they may be morphologically indistinguishable, differ from one another in their power to infect different hosts. Injury to the host may break down its resistance to attack and may render it liable to infection by a species to which it is normally immune.

Under favourable conditions rust appears suddenly, and spreads with great rapidity. Eriksson believed such epidemics to depend on the presence in the seeds or buds of the host plant of the protoplasm of the rust, indistinguishably mingled with that of the host, a mixture to which the term **mycoplasm** was applied. He considered that the protoplasm of the fungus remained unaltered till the leaves were formed; under appropriate conditions it then separated itself rapidly from that of the host and developed into the ordinary spore-bearing mycelium. The investigations of Marshall Ward and others have not substantiated this hypothesis.

**Nuclear Division.** It would appear, from the work of various observers, that nuclear division in the Uredinales has undergone a process of simplification and in some cases it shows but few of the characters of normal mitosis. In the spermatial hyphae of *Gymnosporangium clavariaeforme*, for example, Blackman has described a condensation of the nucleus to form a deeply staining body out of which the nucleolus is squeezed. The chromatin is drawn apart into two apparently homogeneous masses between which a kinoplasmic thread represents the spindle. A similar process takes place in the division of the conjugate nuclei in this and other forms, but the spindle is generally recognizable somewhat earlier, at a time when the chromatin of each nucleus still forms a single mass. As a rule the spindles of the conjugate nuclei lie parallel one to another (fig. 188).

Moreau, following Sappin-Trouffy, has recorded two chromosomes or chromatin masses formed from each nucleus in various Uredineae. Olive on the other hand in *Triphragmidium Ulmariae* and *Uromyces Scirpi* has found a clearly defined spindle and centrosomes and has succeeded in recognizing several separate chromosomes; a similar state of affairs has been recorded by Christman for *Phragmidium speciosum* so that it would appear that the different species of rusts are at dissimilar levels in this matter, though a further study of carefully fixed material might be undertaken with advantage.

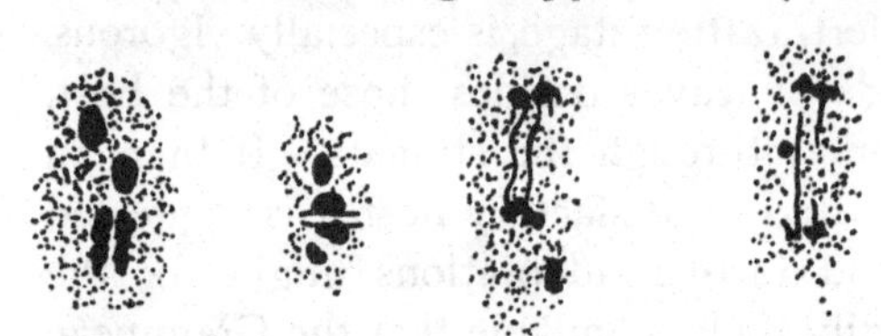

Fig. 188. *Uromyces Poae* Raben.; conjugate divisions in aecidium, × 1330; after Blackman and Fraser.

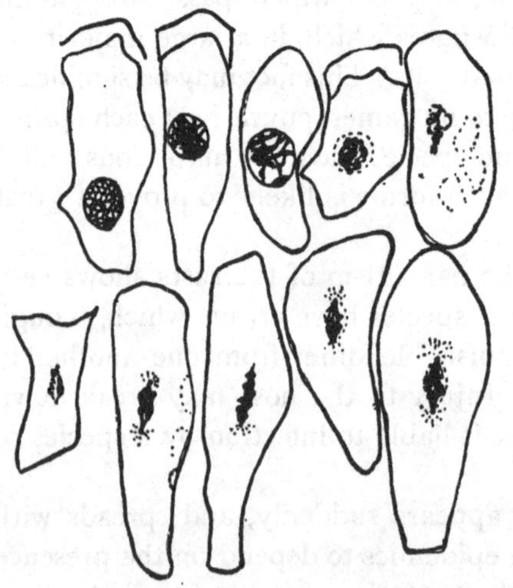

Fig. 189. *Coleosporium Senecionis*; mitosis in teleutospores; after Arnaud.

In all cases, however, the divisions of the fusion nucleus of the teleutospore are much more elaborate than those in the vegetative cells and show some of the characteristics of a meiotic phase. In *Coleosporium* (fig. 189) the fusion nucleus at first possesses a well-marked reticulum of interlacing threads. This undergoes a stage of concentration in one part of the nuclear area, which no doubt corresponds to synapsis, and afterwards loosens out, increases in thickness and forms a spireme. The spireme breaks up and its segments are seen to be double throughout their length. In the meantime centrosomes and spindle fibres have appeared and characteristic gemini are recognizable on the spindle.

Moreau describes only two, but Harper and Holden found a larger number, which became crowded together and more or less fused during the later stages of the first and also during the second division.

In *Gymnosporangium clavariaeforme* (fig. 190) the first division is initiated, as in *Coleosporium*, by a synaptic phase, after which a spireme is formed and breaks up into chromosomes. These pass on to the spindle but soon lose their individuality and travel in irregular masses to the poles.

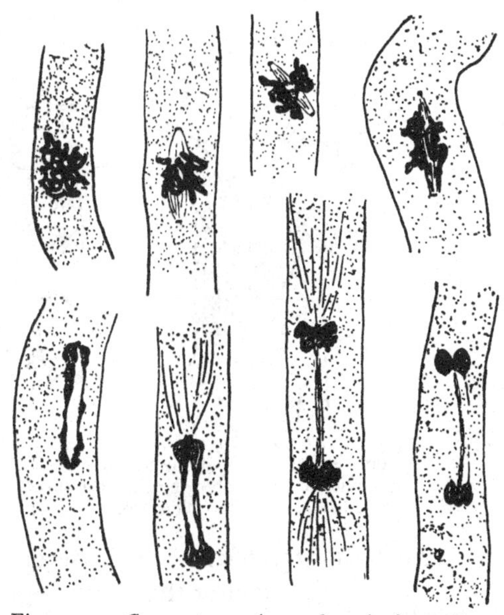

Fig. 190. *Gymnosporangium clavariaeforme* Rees; first division in basidium, × 1460; after Blackman.

The development of the spindle has also been traced in this species; its formation is extra-nuclear and it lies free in the cytoplasm before coming into relation with the dividing nucleus. This type is fairly common among animals but is of exceptional occurrence in plants.

**Nuclear Association.** The cytology of the aecidium was first described in detail in 1904 by Blackman, for *Phragmidium violaceum*, a species occurring on the bramble. The aecidium here is of the caeoma type, consisting of a group of fertile cells of indefinite extent and usually bounded at the periphery by a number of thin-walled paraphyses.

Its formation begins by the massing of hyphae below the epidermis of the leaf where they form a series of uninucleate cells two or three layers thick. The cells next the epidermis increase in size and each divides by a transverse wall parallel to the surface of the leaf, separating an upper sterile cell from the fertile cell below. The sterile cells remain cubical and ultimately disintegrate; the fertile cells elongate to form a more or less regular layer and paired nuclei appear in them, first at the centre and later towards the periphery of the group (fig. 191).

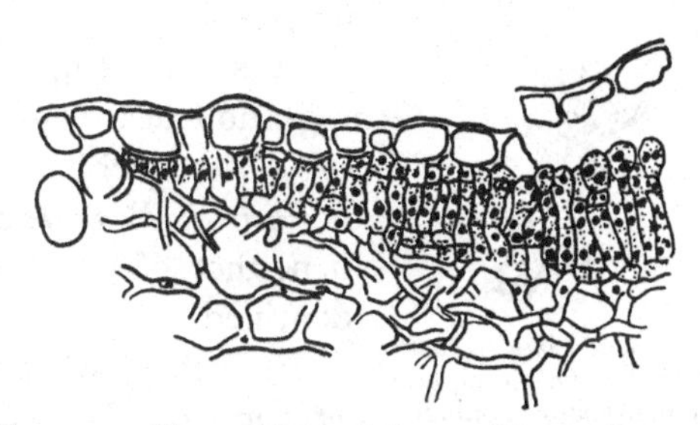

Fig. 191. *Phragmidium violaceum* Went.; caeoma, × 240; after Blackman.

The second nucleus in the

fertile cells of *Phragmidium violaceum* was shown by Blackman and subsequently by Welsford to be derived from one of the smaller cells at the base of the fertile layer. It is thus a vegetative nucleus; it enters the fertile cell by migrating through the wall, becoming much drawn out and laterally compressed. It leaves a pore which may be identified after its passage (fig. 192).

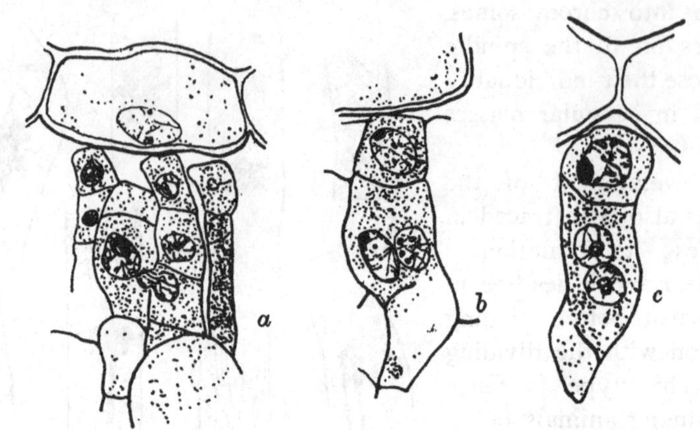


Fig. 192. *Phragmidium violaceum* Went.; caeoma; *a*. migration of nucleus from vegetative cell of one hypha to fertile cell of another, × 1040; *b*. and *c*. binucleate cells showing the pore through which the second nucleus has passed, × 1010; after Welsford.

After entering the fertile cell the second nucleus is at first smaller and denser than that originally present, but soon becomes similar to it in size and consistency. The fertile cell now elongates and in doing so pushes through and destroys the sterile cell above. The associated nuclei divide simultaneously, a transverse wall is formed between the pairs of daughter nuclei cutting off the aecidiospore mother-cell, and the process is several times repeated so that a row of cells is formed. Each of these divides again to separate a small, binucleate intercalary cell below from the binucleate aecidiospore above.

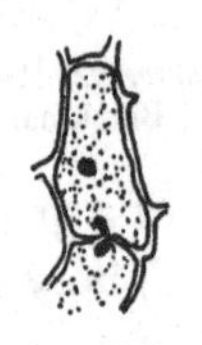

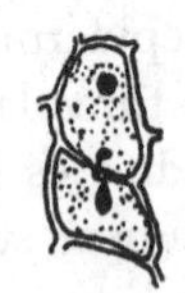

Fig. 193. *Uromyces Poae* Raben.; nuclear migrations in young aecidium × 950; after Blackman and Fraser.

A similar type of development initiated by the migration of a vegetative nucleus into the fertile cell, was observed by Blackman and Fraser in the aecidia of *Puccinia Poarum* and *Uromyces Poae* (fig. 193). But in neither of these was the sterile cell satisfactorily identified.

In such cases it seems reasonably clear that the entrance of the second nucleus is not a primitive process but a form of reduced fertilization where a vegetative nucleus has replaced that of the no longer functional male element. As already shown there is a strong presumption that this male element was the spermatium and the fertile cell may then be regarded as an oogonium and the young aecidium as a group or sorus of female reproductive organs.

In this connection Blackman has suggested a possible origin of the sterile cell; in *Phragmidium violaceum* he found it to be occasionally elongated and pushed up between the cells of the epidermis so that it was covered only by the cuticle (fig. 194); if in the past it broke through this also, it would have formed an efficient trichogyne and may well have functioned as such.

In the related species *Phragmidium speciosum*, Christman, in 1905, described a similar development of sterile and reproductive cells, but in this case the fertile cells become inclined one towards another in pairs and, at the point of contact, the walls dissolve so that the protoplasts come into relation, at first through a small pore, but later along the greater part of their length. Binucleate cells are thus formed (fig. 195), conjugate division

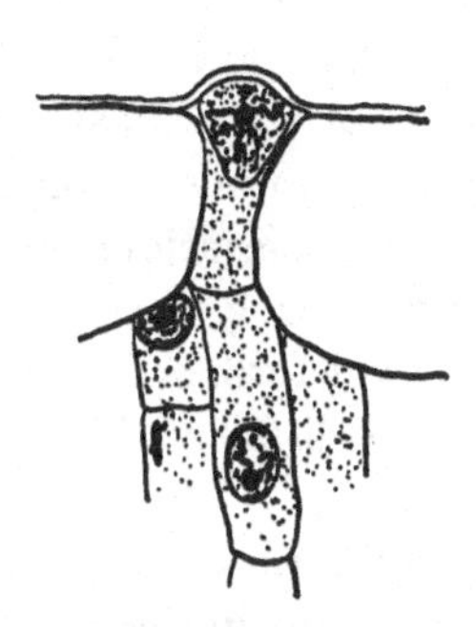

Fig. 194. *Phragmidium violaceum* Went.; caeoma; sterile cell pushing up between epidermal cells of host, ×1300; after Blackman.

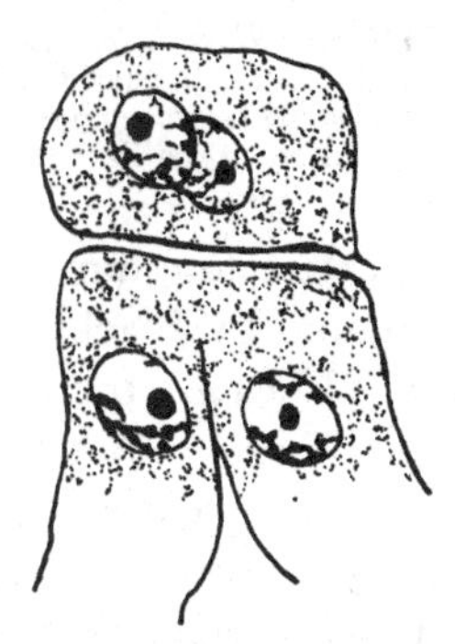

Fig. 195. *Phragmidium speciosum* Fr.; fertile cells after conjugation; aecidiospore mother-cell above; after Christman.

takes place and aecidiospore mother-cells are cut off so that a single row of aecidiospores is developed from each pair of gametes. Christman regards the fertile cells as isogametes between which conjugation takes place, and the sterile cells merely as buffers, of which the function is to assist in the rupture of the epidermis.

His observations on *Phragmidium speciosum* were confirmed in 1906 by Blackman and Fraser for *Melampsora Rostrupi* on *Mercurialis* (fig. 174); these authors pointed out that the fusion of the fertile cells is a reduced fertilization, strictly comparable to that in *Ph. violaceum* but achieved by the union of female cells in pairs instead of by the entrance of a vegetative nucleus into the female cell. This interpretation is confirmed by the fact that Moreau found both processes (cell-fusion being considerably more common than migration) in the same caeoma in *Phragmidium subcorticium*. Since 1905 nuclear association by the fusion of fertile cells in pairs has been observed in a number of species, and seems, according to our present knowledge, to be the usual method.

In the primary uredosorus of *Triphragmidium Ulmariae* and in certain other species, Olive, in 1908, found arrangements of an intermediate type. Here the cells of the young fructification form a more or less regular layer and cut off sterile cells in the usual way. Other hyphae then push up among them, and cell fusion and nuclear association take place (fig. 196). Fusion begins through a narrow pore which afterwards broadens, and in many cases the nucleus of the younger cell migrates into the older one. The process is thus intermediate between those first observed by Blackman and Christman respectively, and the younger hypha, which does not cut off a sterile cell, may be regarded as either a vegetative structure or a gametangium. Olive suggests that the migrations, recorded by Blackman and by Blackman and Fraser, may be merely early stages of a completer cell-fusion; the recent critical work of Welsford, however, negatives this hypothesis, nor would the occurrence of cell-fusion be of much importance once nuclear association had taken place.

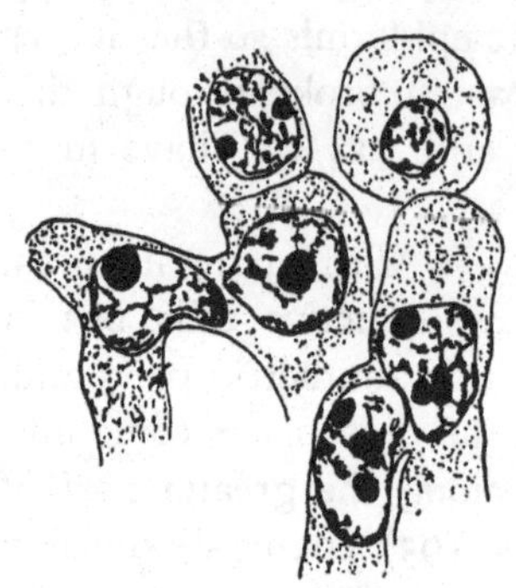

Fig. 196. *Triphragmidium Ulmariae* (Schum.) Link; primary uredosorus; condition intermediate between migration and conjugation of fertile cell; after Olive.

**Phylogeny.** The interpretation given to the processes which take place in the aecidium affects the conception of other spore-forms in the Uredinales and indeed of the phylogeny of the group.

For Christman, the sporophyte arises by the conjugation of undifferentiated or scarcely differentiated isogametes which fuse to form the basal cells of the aecidium. The spermatia cannot on this interpretation be male organs, and he regards them as the once-functional asexual spores of the gametophyte. The basal cells of the aecidium are homologized with those of the uredo- and teleutosori, and the fact is emphasized that the basal cells of the primary uredosorus and sometimes of the teleutosorus also may arise by cell-fusions similar to those in the aecidium. Christman is inclined therefore to regard the *micro*- species, in which the only spore-forms are teleutospores, or teleutospores and spermatia, as the primitive rusts, and to see in them a gametophytic mycelium bearing asexual spores (spermatia) and undifferentiated gametes by the union of which the basal cells of the teleutosorus are produced. Outgrowths of these cells bear the teleutospores in the germination of which the sporophyte comes to an end and the new gametophyte is initiated.

Into this simple life-history the uredospore and aecidiospore are held to be afterwards successively inserted as extensions of the sporophytic phase.

From such a point of view the rust might be related directly to the Phycomycetes or other simple forms.

Blackman's work, on the other hand, indicates that the spermatium is an abortive male element, the fertile cells of the aecidium are female organs which, in the absence of normal fertilization, either fuse in pairs or receive a vegetative nucleus by migration. In *micro-* forms the female organs have disappeared and the abbreviated life-cycle, like that of the pseudapogamous ferns, shows the sporophyte initiated by an association of two vegetative nuclei. The *eu-* forms (or the *-opsis* forms with teleutospores, aecidiospores and spermatia), are therefore primitive and the forms with a shorter life-history are of secondary origin and reduced. The female organ consists of two cells, the upper of which may have functioned as a trichogyne.

Comparing the two hypotheses it may be noted that Blackman's has the advantage of correlating all the known facts, since the association of female nuclei in pairs and of female and vegetative nuclei are both observed methods of replacing normal fertilization. Christman, on the other hand, is obliged to ignore the migrations of vegetative nuclei, or to regard them as pathological. Even for this reason alone it would appear, in the present state of our knowledge, more probable that the Uredinales are a group in which the normal sexual process has disappeared, and is replaced by various forms of pseudapogamy. The young aecidia must then be regarded as groups of female organs, each consisting of a fertile and a sterile cell, and the spermogonia would appear to be corresponding groups of male organs, the spermatia or antheridia, with the filaments which bear them.

A third suggestion proposes *Endophyllum* as a primitive form. The mycelium of uninucleate cells bears spermatia and cluster-cups. At the base of the cup fusion of fertile cells in pairs occurs, and spores and intercalary cells are produced in chains. The spore germinates by the formation of a septate basidium on which four basidiospores are produced in *E. Euphorbiae*, but an irregular number, sometimes as many as eight from one cell, in *E. Sempervivi*.

In *Uromyces Cunninghamianus* (on *Jasminum*) Barclay, in 1891, observed that the aecidiospore germinates by a tube in which one transverse wall is formed and the cells give rise to secondary hyphae which produce infection. On the resultant mycelium, spores similar in arrangement to aecidiospores are formed; so that here, as in *Coleosporium*, we have the accessory spores of the diplophase produced in chains. In the same sorus teleutospores, which are in this genus unicellular, may also arise. The cytology of this species has not been investigated, but there is an indication of a transition between *Endophyllum* and the *eu-* forms It is postulated that the cells of the promycelium of an *Endophyllum*-like ancestor might have produced infection directly instead of by means of basidiospores. Nuclear fusion and meiosis would be postponed till the formation of teleutospores, which would arise as the typical spores of this new vegetative sporophyte. Later the teleutospore would become specialized as a resting spore and uredospores would take on the function of rapid propagation of the plant.

But *Endophyllum* has a definite cluster-cup, and well-marked peridium, and is therefore not likely to be primitive, though it may point to an hypothetical primitive ancestor; a species must rather be sought with a caeoma from the aecidiospores of which promycelia are produced. Such a form has been recognized in *Kunkelia nitens.*

There remain to be considered the various forms of teleutospore. Presumably the unicellular type is more primitive than the multicellular. Simple teleutospores occur in *Melampsora* (where also the aecidiospores are developed in a caeoma), in *Uromyces* and in the Coleosporiaceae. In the last-named family meiosis and septation take place inside the teleutospore wall, but the elaboration of the other spore-forms forbids this group being regarded as primitive, though the internal basidium may be.

It may be hazarded that the ancestral rust bore spermogonia or groups of male organs and groups of female organs of the caeoma type, that the individual male organ was an antheridium or spermatium set free from its parent hypha, that the female organ consisted of a fertile cell or oogonium and a sterile cell which was perhaps elongated to form a trichogyne reaching up to the surface of the host, and that the product of fertilization was a series of aecidiospores. It may further be suggested that either the aecidiospore germinated by giving rise at once to a promycelium or that an alternation of vegetative generations occurred and that the sporophyte bore simple teleutospores or tetrasporangia inside which septation took place.

The members of the Uredinales may be arranged in four families:

| | |
|---|---|
| Germination by a promycelium (except in *Chrysopsora*) | |
| Teleutospores stalked | PUCCINIACEAE. |
| Teleutospores sessile | |
| arranged in series but separating later | CRONARTIACEAE, |
| one or many celled, loose in tissue of host or united in a flat layer under the epidermis | MELAMPSORACEAE. |
| Germination without a promycelium, formation of basidium internal; teleutospores sessile or with a lateral pedicel | COLEOSPORIACEAE. |

### *Pucciniaceae*[1].

The teleutospores of the Pucciniaceae are provided with a stalk which is often well developed, but is in some cases short or becomes detached at an early stage (deciduous). The teleutospores are one-celled in *Uromyces* and *Hemileia*, two-celled in *Puccinia* and *Gymnosporangium*; they are made up of three cells in *Triphragmidium*, and in *Phragmidium* of three or more cells. *Gymnosporangium* is further characterized by the long pedicels of the teleutospores and the fact that they are imbedded in a gelatinous mass. The uredospores are solitary and the aecidiospores produced either in caeomata (*Triphragmidium*, *Phragmidium*), or in aecidia, which in *Gymnosporangium* are commonly elongated to form flask-shaped or cylindrical roestelia.

This family probably includes some of the most highly developed members of the Uredinales, but it includes also several species with caeomata, and one, *Chrysopsora Gynoxidis*, belonging to a monotypic genus in Ecuador, in which the two cells of the stalked teleutospore germinate by internal septation and the protrusion of sterigmata bearing basidiospores as in *Coleosporium*.

### *Cronartiaceae*

In the Cronartiaceae the teleutospores are unicellular and sessile, so that they simulate multicellular spores. In *Chrysomyxa* they form waxy crusts, and in *Cronartium* a cylindrical body. A pseudoperidium is developed around the aecidiospores.

The genus *Endophyllum* is sometimes placed here, sometimes a separate family, the Endophyllaceae, is constituted for it; it differs from the rest of the Cronartiaceae and from the majority of the rusts in the fact that its basidia are developed from spores resembling aecidiospores which arise in an aecidium-like sorus protected by a pseudoperidium.

### *Melampsoraceae*

The teleutospores are sessile, loose in the tissue of the host in *Uredinopsis*, in the other members of the family grouped in a flat layer under the epidermis. In *Melampsora* and its immediate allies they are unicellular; in other genera they are divided either vertically into two cells or by cruciately arranged septa into four.

The aecidiospores may be surrounded by a pseudoperidium or arranged in a caeoma; sometimes a pseudoperidium is present around the uredosorus also.

[1] For Bibliography of this and other families, see the end of the group.

### *Coleosporiaceae*

The outstanding character of the Coleosporiaceae is the method of germination of the unicellular teleutospore which undergoes septation directly and, in *Coleosporium* and *Ochropsora*, without the protrusion of a promycelium; in *Zaghouania* the contents of the teleutospore divide within the teleutospore-wall to form four cells, but emerge before the basidiospores appear.

The aecidia are cup-shaped in *Ochropsora*, but in *Coleosporium* and *Zaghouania* they are of the peridermium type with a cylindrical, more or less inflated peridium; this elaborate type of aecidial sorus makes it impossible to regard the family as primitive though it may perhaps have branched off early from the line leading to the commoner type of rust.

## UREDINALES: BIBLIOGRAPHY

1889 PLOWRIGHT, C. B. A Monograph of the British Uredineae and Ustilagineae. Kegan Paul, Trench & Co., London.

1891 BARCLAY, A. On the Life history of a remarkable Uredine on *Jasminum grandiflorum*. Trans. Linn. Soc. Bot. ii, p. 141.

1895 POIRAULT, G. and RACIBORSKI, M. Sur les noyaux des Urédinées. Journ. de Bot. ix, pp. 318 and 381.

1896 SAPPIN-TROUFFY, P. Recherches histologiques sur la famille des Urédinées. Le Botaniste, v, p. 59.

1902 DUMEE, P. and MAIRE, R. Remarques sur le *Zaghouania Phyllyreae*, Pat. Bull. Soc. Myc. France, xviii, p. 17.

1903 HOLDEN, R. J. and HARPER, R. A. Nuclear Divisions and Nuclear Fusion in *Coleosporium Sonchi-arvensis*, Lév. Trans. Wisconsin Acad. Sci., Arts and Letters, xiv, pt. i, p. 63.

1904 BLACKMAN, V. H. On the Fertilization, Alternation of Generations and general Cytology of the Uredineae. Ann. Bot. xviii, p. 323.

1904 TRANZSCHEL, W. Ueber die Möglichkeit die Biologie wertwechselnder Rostpilze auf Grund morphologischer Merkmale vorauszusehen. Arb. Kais. Petersburg Naturf. Gesell. xxxv, p. 1.

1905 CHRISTMAN, A. H. Sexual Reproduction in the Rusts. Bot. Gaz. xxxix, p. 267.

1905 ERIKSSON, J. On the Vegetative Life of some Uredineae. Ann. Bot. xix, p. 53.

1905 WARD, H. MARSHALL. Recent Researches on the Parasitism of Fungi. Ann. Bot. xix, p. 1.

1906 BLACKMAN, V. H. and FRASER, H. C. I. Further Studies on the Sexuality of the Uredineae. Ann. Bot. xx, p. 35.

1906 McALPINE, D. The Rusts of Australia. J. Kemp, Melbourne.

1907 CHRISTMAN, A. H. The Nature and Development of the Primary Uredospore. Trans. Wisconsin Acad. Sci. xv, p. 517.

1907 CHRISTMAN, A. H. Alternation of Generations and the Morphology of the Spore-forms in Rusts. Bot. Gaz. xliv, p. 81.

1908 OLIVE, E. W. Sexual Cell-Fusions and Vegetative Nuclear Divisions in the Rusts. Ann. Bot. xxii, p. 331.

1910 DITTSCHLAG, E. Zur Kenntnis der Kernverhältnisse von *Puccinia Falcariae*. Centralbl. f. Bakt. Abt. ii, B. 28.

1911 HOFFMANN, A. W. H. Zur Entwickelungsgeschichte von *Endophyllum Sempervivi.* Centralbl. für Bakt. Parasit. Infect. xxxii, p. 137.
1911 MAIRE, R. La Biologie des Urédinales. Prog. Rei Bot. iv, p. 109.
1911 OLIVE, E. W. Origin of Heteroecism in Rusts. Phytopathology i, p. 139.
1911 SHARP, L. W. Nuclear Phenomena in *Puccinia Podophylli.* Bot. Gaz. li, p. 463.
1912 FROMME, F. D. Sexual Fusions and Spore Development of the Flax Rust. Bull. Torrey Bot. Club, xxxix, p. 113.
1912 RAMSBOTTOM, J. Some Notes on the History of the Classification of the Uredinales with full list of British Uredinales. Brit. Myc. Soc. IV, p. 77.
1912 WERTH, E. and LUDWIG, K. Zur Sporenbildung bei Rost- und Brandpilzen. Ber. deutsch. Bot. Ges. xxx, p. 522.
1913 ARNAUD, G. La Mitose chez *Capnodium meridionale* et chez *Coleosporium Senecionis.* Bull. Soc. Myc. de France, xxix, p. 345.
1913 GROVE, W. B. The British Rust Fungi. Camb. Univ. Press.
1913 GROVE, W. B. The Evolution of the Higher Uredineae. New Phyt. xii, p. 89.
1914 FROMME, F. D. The Morphology and Cytology of the Aecidium Cup. Bot. Gaz. lviii, p. 1.
1914 KUNKEL, L. O. Nuclear Behaviour in the promycelia of *Caeoma nitens,* Burrill, and *Puccinia Peckiana* Howe. Am. Journ. Bot. i, p. 37.
1914 KURASSANOW, L. Über die Peridienentwicklung im Aecidium. Ber. deutsch. Bot. Ges. xxxii, p. 317.
1914 MOREAU, Mme F. Les phénomènes de la sexualité chez les Urédinées. Le Botaniste, xiii, p. 145.
1915 WELSFORD, E. J. Nuclear Migrations in *Phragmidium violaceum.* Ann. Bot. xxix, p. 293.
1916 KUNKEL, L. O. Further Studies on the orange rusts of *Rubus* in the United States. Bull. Torrey Bot. Club, xliii, p. 559.
1917 ARTHUR, J. C. Orange Rusts of *Rubus.* Bot. Gaz. lxiii, p. 501.

# GENERAL BIBLIOGRAPHY


1861 TULASNE, R. C. and L. Fungi Hypogaei. Klincksieck, Paris.

1861–5 TULASNE, R. C. and L. Selecta Fungorum Carpologia. Imperial. Typograph., Paris.

1882–1913 SACCARDO, P. A. Sylloge Fungorum. Published by the author. Padua.

1884 DE BARY, A. Comparative Morphology and Biology of the Fungi, Mycetozoa and Bacteria. Eng. Trans. 1887. Clarendon Press, Oxford.

1884–97 RABENHORST, L. Kryptogamen Flora. Pilze, by G. WINTER and E. FISCHER. Ed. Krummer, Leipzig.

1887 PHILLIPS, W. A Manual of British Discomycetes. Kegan Paul, Trench & Co., London.

1892 VON TAVEL, F. Vergleichende Morphologie der Pilze. Fischer, Jena.

1892–5 MASSEE, G. British Fungus Flora. Bell & Sons, London.

1895 VON TUBEUF, K. F. Diseases of Plants. Eng. Trans., 1897. Longmans, Green & Co., London.

1897–1900 ENGLER, A. and PRANTL, K. Die natürlichen Pflanzenfamilien. Fungi, by J. SCHRÖTER, G. LINDAU, E. FISCHER and P. DIETEL. Engelmann, Leipzig.

1904–11 BOUDIER, E. Icones Mycologicae. Klincksieck, Paris.

1906 MASSEE, G. A Text Book of Fungi. Duckworth & Co., London.

1909 BULLER, A. H. R. Researches on Fungi. Longmans, Green & Co., London.

1909 SWANTON, E. W. Fungi and How to Know Them. Methuen, London.

1910 DUGGAR, B. M. Fungous Diseases of Plants. Ginn & Co., New York.

1913 MASSEE, G. Mildews, Rusts and Smuts. Dulau & Co., London.

1913 STEVENS, F. L. The Fungi which cause Plant Disease. Macmillan Co., New York.

1918 BUTLER, E. J. Fungi and the Diseases of Plants. Thacker, Spink & Co., Calcutta.

1918 HARSHBERGER, J. W. A Textbook of Mycology and Plant Pathology. J. & A. Churchill, London.

1919 HILEY, W. E. The Fungal Diseases of the Common Larch. Clarendon Press, Oxford.

# INDEX

*A glossary has not been prepared for this volume, but the page on which the definition of a technical term will be found is shown in the index in clarendon type, and the same method is used for indicating the principal reference to a family or genus.*

PRINTED IN ENGLAND BY J. B. PEACE, M.A.
AT THE CAMBRIDGE UNIVERSITY PRESS

Printed in the United States
By Bookmasters